MODELING THE EUTROPHICATION PROCESS

ESEA
TITLE-II GRANT

MODELING THE EUTROPHICATION PROCESS

Edited by

E. JOE MIDDLEBROOKS
Professor and Head, Division of Environmental Engi-
neering, College of Engineering, Utah State University,
Logan, Utah.

DONNA H. FALKENBORG
Editor, Utah Water Research Laboratory, Utah State
University, Logan, Utah.

THOMAS E. MALONEY
Chief, National Eutrophication Research Program, Na-
tional Environmental Research Center, Environmental
Protection Agency, Corvallis, Oregon.

ann arbor science PUBLISHERS INC.
POST OFFICE BOX 1425 • ANN ARBOR, MICHIGAN 48106

Library of Congress Catalog Card No. 74-76411
ISBN 0-250-40059-6

Manufactured in the United States of America

PREFACE

The papers presented in this monograph are original contributions collected from a group of scientists and engineers actively engaged in the modeling and evaluation of the eutrophication process. The group was convened at Utah State University, Logan, Utah, with the support of the U.S. Environmental Protection Agency, for the purpose of exchanging ideas and the most recent information available pertaining to the eutrophication process, and to keep one another aware of ongoing research in this area.

This work was partially supported by Grant No. R-802462 received from the National Eutrophication Research Program of the Environmental Protection Agency and the Utah Water Research Laboratory, Utah State University.

The great expenditure of time and effort made by the authors of the papers is gratefully acknowledged. Without such willingness to share knowledge and experiences, publications such as this would be impossible.

TABLE OF CONTENTS

Page

TABLE OF CONTENTS (CONTINUED)

Page

SYNOPSIS OF WORKSHOP ON MODELING OF THE

EUTROPHICATION PROCESS

P. H. McGauhey[1]

There are two words in our modern vocabulary that always scare the stuffing out of me, to the extent that it takes a great deal of intellectual aspirin to reassure me again. One of these is "innovation"—the other is "modeling." Invention and ingenuity bear no resemblance to what I associate with innovation. To me "innovation" might be defined as a "vague and formless vapor which prevades the dreams of the impractical or the incompetent fellow."

But why should I be frightened by the bilious dreams of others? My own are more than I can cope with. The answer is that experience has led me to expect that embracing innovation means that the meager funds allocated to doing something useful will now be siphoned off to support some "new innovative program" based on, to paraphrase Dr. Krause of Harper's magazine —"the systematic misuse of jargon especially invented for that purpose." The result is that by the time the new program dries up both itself and the source of funds, it is impossible to say whether the innovator has rediscovered the wheel, proved the wheel impossible, or has simply retitled his original madness and submitted it as a Final Report.

Sometimes innovation is to be implemented by the innovator by the use of a model. This prospect triggers within me a sequence of hallucinations which might well be called a "bad trip." First I see a set of rectangles, somewhat like a draftsman's representation of the buildings on a university campus. These are connected by lines which could well represent those used by the architect to indicate paved walks, but never used by students in going from class to class. This disparity indicates that the model is non-linear. The variety here is little short of infinite. Nevertheless this is a static model suited only to the campus at one lost instant in time.

To make the model useful for year 2020 planning, it must be made dynamic. This means that it must be made to take into account such factors as the construction of a new building, moving the English department to the

physics building, the introduction of innovative programs, and, yes, the disappearance of some vestigical curricula.

By this time I have become a little panicky. But panic has only begun. The model must now be programmed for the computer. When this is done, it is alleged, the model will be universalized—and available for a fee wherever universities have buildings and student paths.

At this point in my dreams I imagine myself getting a consulting job, say at BYU, expecting to sub-contract the model work to the owners of the USU model. Unfortunately I get the job. Only then do I learn that the model will have to be modified somewhat because BYU has a different spectrum of buildings and their juxtaposition is different than that of similar buildings at USU. In fact, the only similarity between the two campuses is that students move in straight lines unrelated to paved walkways.

To make the necessary adjustment great cost is required in "expanding the model"—which is what its originators are really interested in anyway. Ultimately, the model diverges in a vain effort to include all pertinent variables instead of converging on the BYU situation; the cost is astronomical; I go broke—and wake up in a cold sweat to thank goodness that once again the whole thing was but a dream.

Sometimes these nightmares include a program which is full of big Sigmas; little i,j's, and dots leading off toward the horizon or down into the pit where little "n" resides. This model is based upon converting the things I don't know about the problem into mathematical symbols. Then I find that I can use the model only by calibrating it. This, in turn, is possible only by substituting for the symbols the information I didn't have in the first place. Again I go broke and wake up in a lather.

But why do I tell you these things today as I set out to speak of the workshop's accomplishments? In fact, why did I agree to discuss so frightening a subject in the first place? The principal answer is that I always like to come to Logan. Of lesser importance is the fact that I have to live with this type of model, and know just enough about eutrophication to believe that if we are ever

---

[1]P. H. McGauhey is Professor Emeritus, Sanitary Engineering Research Laboratory, University of California, Berkeley, California.

going to be able to predict the integral of its complex relationships we are going to have to learn how to model the eutrophication process. So I concluded in advance of any familiarity with the workshop papers that it couldn't be all bad. And indeed that has turned out to be the case.

As a result of the foreshortening of time, which is bedeviling us all, only two of the papers arrived in advance of the workshop, hence I must rely on notes taken in the darkness of our slide-ridden environment here to recapture my impressions of what was said by the speakers.

All of you, plus several who had the good fortune to escape before this hour, heard the presentations and took such exception as came to mind at the moment. Therefore, I shall not repeat from imperfect memory what you have heard.

Rather I shall give my hastily prepared impression of what the workshop brought into sharp focus—asking you to bear in mind my opening confession and to weigh the reality of the workshop against my fears.

Several important items impressed me:
1. The first is that young men and women are attacking with both energy and enthusiasm the difficult problem of modeling the response of ecosystems to limnological phenomena.
2. The fact that this attack is involving a wide spectrum of disciplines simultaneously—not just the mathematician searching for a place to use his systems approach, the biologist or other scientist in need of witchcraft, or the technologist seeking a way to circumvent the lasar-beamed eye of the self-anointed environmentalist. Although this item was not explicitly discussed, it seemed to me to be implicit in the presentations and in a general mood of sanity which prevaded the workshop.
3. I sensed an undercurrent which I find to be healthy, although not necessarily past the time when abortion is feasible. This is the prospect of society once again giving appropriate recognition to the role of research. I detect this in both:
   a) The willingness of people to press on with their research in spite of reduced support from public agencies.
   b) The willingness of representatives of public agencies to encourage continued research despite the heavy hand laid upon them from above.

Of all matters of broad import coming out of this workshop, none is more clearly evident than the need for sophisticated systems toward which those reporting are groping. This may be documented in two contexts:
1. The needs of EPA in fulfilling its responsibility under existing law.

2. The potential of the present state-of-the-art to fill those needs.

Without going afield to express my personal unhappiness with PL92-500 and the phenomenon of dictatorship by administrative fiat, it is clearly evident that EPA badly needs, and wants, a lake model which with obtainable inputs in any specific lake situation can be employed to predict the response of that lake to its various influents, and hence to disclose the necessary controls and monitoring programs for protecting the lake.

The workshop revealed that the present state-of-the-art is far too primitive to produce either the universal model or to evaluate correctly the inputs which should go into it.

I think it was clear to everyone participating in this workshop that in our ability to model a limnological situation we are as infants newly discovering that modeling clay will "squish" in our hands.

Our inability to construct or to formulate the universal model leads to the next best approach; that is to make specific models which may be used to protect specific lakes; and then try to expand these models to universalize them. It was fully recognized, and specifically stated in the discussion, that we cannot hope to deal with all lakes by the specific approach, but some critical local situations may be salvaged. Concerning the subject of the workshop itself, "Modeling of the Eutrophication Process," we did not quibble over definition. In many ways this is good because the last resort of scoundrels is to demand a precise definition of what we are talking about. This quickly reduces any discussion to futility.

It was made clear at the start that we would be discussing ongoing research. It then soon became evident that many of the projects were initiated when eutrophication was defined as the limnological evils wrought by phosphorus; the evils themselves appeared blue-green. Thus definition had therefore been dealt with some time in the past when the projects were conceived.

It was only when we got to the subject of modeling the ecological system and the mathematicians beat the biologists to available phosphorus that the matter of definition arose. And had we gone on for another hour last night we might have had to retitle the workshop. At this point the emphasis of the workshop shifted from the specific case to be universalized, to the rationally pure universal model and how it might be applied to the specific case; basically by setting the tough terms in the model equal to 0.

But, going back to the first papers presented; having previously identified phosphorus as the principal villain much of the discussion centered on making a model which would track down phosphorus through the physical and biological systems of a lake. This I do not deplore,

2

however, for two reasons: 1. In the time which necessarily elapses between identifying a villain and in bringing him to justice through research, greater public concern may have arisen over a whole sequence of new villains. This often leaves us researchers pursuing yesterday's tiger when we ought to be writing proposals to hunt today's. But this is not all bad, because of: 2. In tracking down phosphorus the speakers became more fully aware of the problem of many of its associates; and of the presence of rival groups of villains.

I think the speakers clearly showed their recognition of these many factors which have to be evaluated for sensitivity in natural systems and reckoned with in the ultimate model.

In fact many speakers made liberal use of qualifiers. "This *will vary*," they said "depending upon ... etc." They then listed a few of the many influencing factors and hurried on to tell their story without dwelling upon the obscure ways in which this variation takes place.

Some speakers used such terms as "it seems that" without identifying whether the basis of this seeming was a postulate within the intellect, or one possible explanation of disparities between the fictions of nature and the realities of mathematical models.

There were indications, and quite strong ones, in the tone of the workshop, that models and their handmaiden (the solid waste generating computer) are tools for pounding out facts and artifacts that man wants for reasons of environmental control. Their limitations were deplored but no one came right out and admitted that they are at present only in the Stone Age of development as tools. There was still, however, some reverence for the mystique of models. Thus indicating, more from habit than from the arrogance of a few years ago, that model worshipers have not yet all deserted the god of mathematics.

Bierman, after almost making the grievous error of not including this reviewer among his references, rallied to state quite positively that it is not necessary to have a detailed knowledge of metabolic processes to develop a model. This I applaud as important and good news. Some of us older codgers, who were young when probability was as new as models are today, used to apply that tool to the question: What is the probability (or chance) that whatever goes on in all our areas of ignorance will combine in such a way as to produce a particular result.

This result was, of course, always bad—an unacceptable risk, a design loading, or some other disaster or marginal catastrophy. I do not recall that we ever contemplated any beneficial or desirable combinations—only disasters.

This same disaster climate prevails today in relation to water quality. Who ever heard of nutrients, or heat, or

primary productivity being other than bad, or of fish that died of old age, or of algal blooms that failed to affront the eye of the same slobs who litter the lakeshore and the roads leading to it.

One fascinating aspect of the workshop was the evidence that Monod is not dead. Perhaps I am over-sensitive to this possibility because I have long questioned the potential of Monod kinetics to represent the response of systems in which biota get energy from solar sources directly and exercise a considerable number of options in how best to utilize that energy in relation to available nutrients. Therefore I was happy with statements by Bierman et al. in support of this concept.

Evidently, the familiar shelter of Monod still gives some assurance to the venturesome modeler; at least several speakers took comfort in finding a Monod-type expression in their mathematical model. Not that kinetics are not an important factor in simulation, but rather that some men can keep occupied for a lifetime trying to make the response of a lake substantiate the Monod kinetics.

Related to this phenomenon is the most interesting questions raised yesterday: What is the fate of the individual model? Does anyone ever use it? Why is it not further used?

I would suggest that the normal fate of the individual model is to be published. It is not then further used because, as one participant in the workshop suggested, its author was trying to substantiate a different hypothesis than the next author. Why is it not further used may well be that:

1. Its author suffered an energy crisis before it was developed into a universal model.
2. It hasn't the capability of becoming a universal model.
3. Each researcher is forced to work with a simplified model and so omits these factors thought to be most important by the next modeler.
4. The time required to perfect the model exceeds the time period over which support can be generated; i.e. longer than the attention span of society to any one point of view regarding priority of crises.

Items stressed by several speakers or discussants include:

1. Vertical mixing is important.
2. Models should include the effects of sediments in supplying nutrients to overlying waters.
3. Relative recovery rate of a lake may be a key to where measures should be given priority—in the absence of a universal model.
4. It is conceptually feasible to link hydrologic and lake response in a simulating model, or to

relate physical system to the response of biota systems (not vice-versa).

5. Models which link land and water are needed. Activity on land in many cases dominates the input of nutrients.

6. Temperature is a more important factor than we often realize.

7. Feedback of nutrients during growing season is not included in models. This complicates the problem of interpreting the relationship (or meaning) of early fast growing organisms followed by slower growing organisms.

Going back now to the question of definition and of tracking phosphorus, I will resort to a brief commercial and suggest that everyone read the results of EPA-sponsored studies at Lake Tahoe and its associated Indian Creek Reservoir [EPA Report Number 16010-DNY07-(71)]. It raises what I believe is the most important environmental engineering question of our time: HAVING DECIDED WHAT CHARACTERISTICS WE WANT IN A BODY OF WATER, HOW CAN WE PROCEED TO ACHIEVE THEM BY PROCESSING WASTE-WATER AND APPLYING ENVIRONMENTAL CONTROLS? One thing is certain, our research (at ICR, if not elsewhere) shows that any resemblance between the behavior of an impoundment and the quality of its influent water as measured by laboratory analyses continues to be unrevealed. Meanwhile we pile process on process in the forlorn hope that by producing holy water we will do nature an ecological favor.

What we have learned at this workshop is both important and encouraging, yet it seems clear that our destiny is to go "back to the modeling clay" before our objectives can be attained.

4

LONG-TERM LAKE RECOVERY RELATED TO AVAILABLE PHOSPHORUS [1]

E. B. Welch, C. A. Rock, and J. D. Krull [2]

Introduction

Adequate data to predict lake response from large scale manipulations of nutrient budgets are still wanting. A few cases exist in Europe (Mathiesen, 1971; Andersson et al., 1973) which give the impression that nutrient diversion from shallow highly eutrophic lakes is not apt to result in rapid or extensive recovery. Effort in the U.S. is presently underway to reduce the nutrient income to probably 35 to 40 eutrophic lakes (Ketelle and Uttormark, 1971). For most of these projects it is still too early for contributions to the present knowledge of lake response to nutrient income manipulation.

A notable exception is the case of Lake Washington in Seattle, which Edmondson (1956, 1969, 1970, 1972) has thoroughly documented through its enrichment and starvation over the past quarter of a century. This work has not only provided the most positive encouragement for possibilities of lake recovery, but also some of the most convincing evidence of the role of phosphorus as a long-term controlling nutrient. However, Lake Washington is relatively deep (mean depth 33 m), it flushes rather frequently (once per three years), it was never enriched to the point that the hypolimnion became anaerobic, and the major portion of the phosphorus income (60-75 percent) through treated sewage was diverted (Edmondson, 1972).

An interesting contrast to the Lake Washington case is presented by Lake Sammamish which is about one fourth the area of Lake Washington and lies only 20 km to the east. Sammamish is about one half as deep (18 m mean depth), flushes about as frequently (once in 2.5 years), received about the same income of phosphorus before diversion, its hypolimnion does go anaerobic each summer for morphometric reasons, and a much smaller fraction of its phosphorus income (one third) was diverted. The lack of any noticeable response in Lake Sammamish during five years since diversion is puzzling. However, these contrasting conditions between the two lakes has led to some new hypotheses about internal mechanisms that may affect the response–phosphorus income relationship and retard recovery and useful modifications to existing gross phosphorus response models to predict extent and rate of response.

Phosphorus Income

The best estimate to date of the fraction of phosphorus diverted from Lake Sammamish is about one third. A study of waste streams to determine absolute amounts diverted showed a diversion of about 30 percent. Consideration of population equivalents gave 39 percent. Annual monitoring of Issaquah Creek, which is the principal tributary and contributes about 70 percent of the lake's phosphorus, along with 12 minor tributaries, showed a change from a little less than 1 g P/m^2 yr to about 0.6 g P/m^2 yr–a reduction of about 35 percent.

These measurements of surface income may suffer from year-to-year variations on the order of the one third believed to have been diverted. This is because at high flows, above 500 cfs, in Issaquah Creek phosphorus concentration becomes porportional with flow probably because of scouring and erosive activity in the watershed and streambed increasing particulate phosphorus. Thus, high flow years will show increased income compared to low flow years irrespective of the waste diverted. If wastewater into Lake Sammamish had not been intercepted in 1968, the present day waste loading would be five times the pre-diversion level, or a two to three times increase in total income.

The Lake Washington phosphorus income (Edmondson, 1969) was decreased from a rate (about 1.2 g P/m^2 yr) somewhat greater than the pre-diversion income to Lake Sammamish, to a similar post-diversion rate (about 0.5 g P/m^2 yr). Edmondson (1972) has since suggested that the pre-diversion phosphorus income to Lake Washington was actually greater. Clearly then, while a major fraction of the phosphorus income was diverted from Lake Washington only a minor fraction was diverted from Lake Sammamish. Other factors being equal, the extent of response should have been greatest in Washington, but some change would have also been expected in Sammamish.

Lake Response

In contrast to a 70 percent decrease in mean winter phosphorus content in Lake Washington subsequent to

[1] This work was supported in part by EPA research grant No. R-800512 and in part by the National Science Foundation grant No. GB-36810X to the Coniferous Forest Biome, Ecosystem Analysis Studies, US/IBP. This is contribution no. 89 from Coniferous Forest Biome.

[2] E. B. Welch, C.A. Rock, and J.D. Krull are with the Department of Civil Engineering, University of Washington, Seattle, Washington.

diversion, no significant change has been noted in Lake Sammamish phosphorus content. The content in Washington dropped from 63 to 20 µg/l (Edmondson, 1970) while in Sammamish it has varied from 24 to 38 µg/l. As a consequence, spring-summer chlorophyll *a* in Lake Washington decreased from 28 to 8 µg/l (Edmondson, personal communication), while spring-summer chlorophyll *a* in Lake Sammamish has remained relatively unchanged ranging from 5.5 to 8 µg/l. Accordingly, Lake Washington has become much clearer; Secchi disk depths have increased from a summer mean of 1 to 2.5 m (Edmondson, 1972). Secchi disk depth has remained relatively unchanged in Sammamish ranging from summer means of 3.1 to 3.8 m. Maximum Secchi depths have been greatest in the past two years in Lake Sammamish, however.

The pattern of seasonal change in phytoplankton biomass in the two lakes is somewhat different. The large persistent biomass throughout the summer in Lake Washington before diversion is characteristic of enriched lakes (Figure 1). Lake Sammamish, on the other hand has exhibited two patterns of seasonal development, one gradual and somewhat persistent and one abrupt with large maximums followed by a rapid collapse (Figure 1). The differences between the lakes illustrate an effect on pre-growing season nutrient supply. Whether the development is rapid or gradual in Lake Sammamish the average quantity attained is dependent upon the quantity of controlling nutrient initially available. As illustrated in Figure 2, the average growing season biomass attained is reasonably proportional to the initial quantity of phosphorus in four lakes of the Cedar River watershed. As noted, the quantity of phosphorus and chlorophyll *a* have proportionately decreased in Lake Washington since diversion (Edmondson, 1970). The seasonal pattern of chlorophyll *a* in Lake Washington has now approached that of the 1970 and 1972 years in Lake Sammamish.

Hypotheses for Lake Response

The differing response of these two somewhat similar lakes may offer hypotheses for the response to phosphorus income change. Vollenweider (personal com-

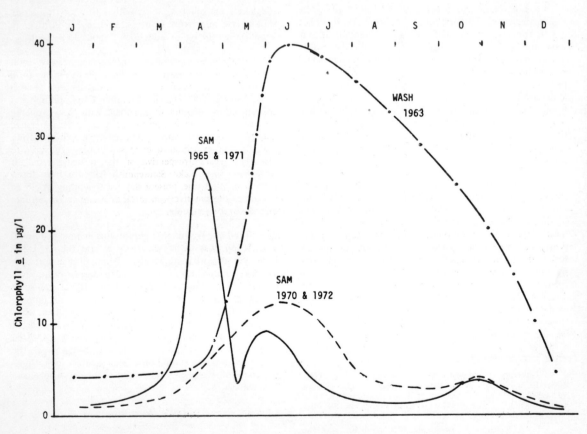

Figure 1. Generalized seasonal patterns of chlorophyll *a* content in Lakes Washington and Sammamish during the years specified. Lake Washington data from Edmondson (1969).

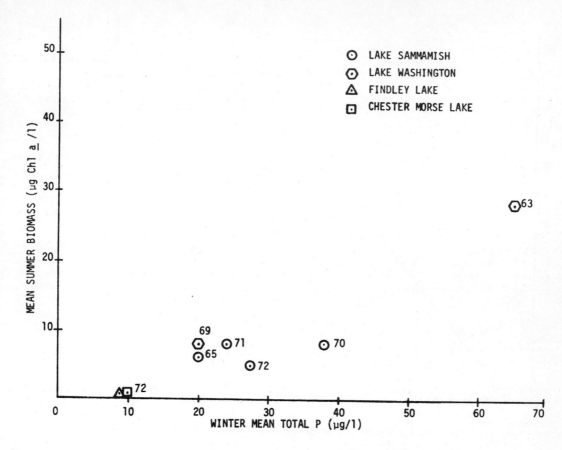

Figure 2. Spring-summer phytoplankton biomass (chl *a*) versus winter phosphorus content in lakes of the Cedar River watershed, Washington, for the years indicated. Lake Washington data from Edmondson (1972).

munication) has suggested that response is probably not linearly related to enrichment and for a lake to show a significant response to a change in nutrient income may depend upon where existing conditions are on the response curve. Figure 3 shows two values for Lake Washington for pre- and post-diversion and one for Moses Lake (Bush and Welch, 1972). One might logically expect that continued enrichment of Lake Washington above the 1963 level would at some point cease to produce a proportional increase in chlorophyll *a*—no doubt long before the light limited hypereutrophic state of Moses Lake is reached. This implies that a similar fractional removal of phosphorus from Moses Lake as occurred for Lake Washington would probably not result in a very dramatic response, which would seem intuitively reasonable.

Results from the phosphorus decrease into Lake Sammamish obviously do not apply to such a logical response curve. In fact, Lake Sammamish is presently receiving as much phosphorus annually as Lake Wash-

ington did before diversion, but produces far less phytoplankton. The current explanation of the lack of response in Lake Sammamish to nutrient diversion is shown by the hypothetical response curve in Figure 3. Because lake plankton chlorophyll *a* has not changed, the decrease in phosphorus income suggests that Lake Sammamish has a more gradual response curve than Lake Washington and that pre- as well as post-diversion phosphorus income rates are on the asymptote of that curve (Figure 3). Accordingly, the income to Sammamish would have to be reduced greatly, in excess of one third, in order to cause a decrease in chlorophyll content. The more gradual response to nutrient income in Lake Sammamish suggests that some internal factor(s) is controlling the availability of the incoming phosphorus to the phytoplankton. One would expect that if phosphorus income were increased further, and there is reason to believe that in lieu of diversion it would now be more than double the pre-diversion income, that chlorophyll *a* would increase at a rate approaching that predicted by the Lake Washington curve.

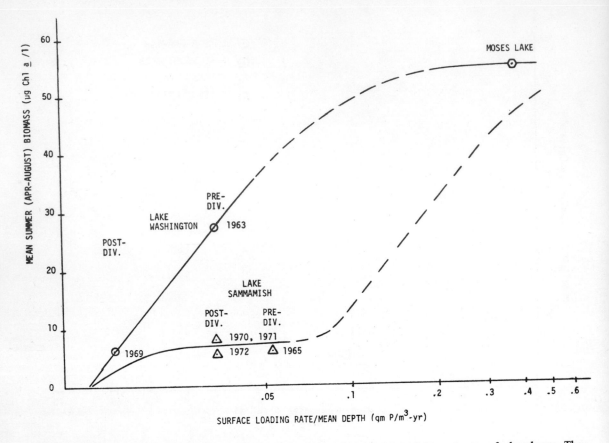

Figure 3. Hypothetical curves for plankton biomass (chl a) in response to annual income rates of phosphorus. The stepped response in Lake Sammamish hypothesized to be controlled by iron. Lake Washington data from Edmondson (1972).

Two areas of evidence support such a hypothetical response curve for Lake Sammamish. The first is that while the reduction in phosphorus supply to Lake Washington resulted in a proportional decrease in the resultant lake P concentration, according to a relationship presented by Vollenweider (1968), the change in supply to Sammamish had no effect on the lake concentration (Figure 4). Since summer phytoplankton biomass is dependent upon the pre-growing season phosphorus content (Figure 2), it is not surprising then that biomass in the lake has not responded to phosphorus income change.

The second area of evidence is that the lake phosphorus content in Sammamish is controlled by iron. Horton (1972) has shown that total iron is closely correlated with total phosphorus as oxygen is exhausted in the hypolimnion during late summer (Figure 5). Although phosphorus increases in the surface waters following lake turnover in late November, it is rapidly complexed by what are probably ferric hydroxides and is resedimented before phytoplankton have a chance to use it when light is adequate in April and May.

Some comparative points between the two lakes which support the strong control that iron has on P content in Sammamish are as follows: Because Lake Washington remains aerobic in the stratified period, iron content only reaches one tenth (Shapiro et al., 1971) of the anaerobic Sammamish maximum. Also, nearly all of the winter phosphorus in Washington is soluble (Edmondson, 1972) while only about one fourth of the total in Sammamish is soluble (ortho).

Thus the internal factor controlling the lake content of P in Sammamish is iron released in large concentrations because of hypolimnetic oxygen exhaustion which subsequently complexes and resediments the large quantities of regenerated phosphorus, greatly limiting its availability

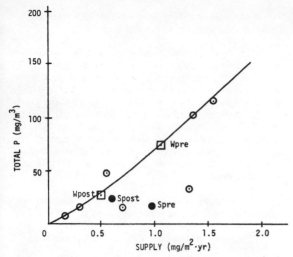

Figure 4. Mean spring or winter (monomictic Lakes Sammamish and Washington) phosphorus concentration as a function of total phosphorus supply after Vollenweider (1969).

to phytoplankton. Since Lake Washington does not go anaerobic the sediments do not give up as much iron and most of the phosphorus is left in the soluble state until removal by phytoplankton in the spring. If P income were increased to Lake Sammamish then some point should be reached where the iron binding capacity is overcome and an increasing portion of the added P would remain available for the vernal phytoplankton growth.

There is some reason to believe that the trophic state of Lake Sammamish has not changed greatly in the past 60 years and therefore has responded little to what has been a considerable increase in nutrient income. A survey in August 1913 showed an oxygen profile similar to that of today (Kemmerer et al., 1924)—about 47 percent saturated at 19 m. Unfortunately the full depth of the water column was not sampled. The Secchi disk depth was 3.3 m, within the range of modern day readings, and the most abundant net plankton algae were *Microcystis* and *Fragilaria* with *Anabaena, Aphanocapsa, Asterionella* and *Cyclotella* of less but about equal abundance. Again this is the present picture in the lake for that time of year. Confirmation of the degree of change in trophic state must await more thorough sediment core analysis.

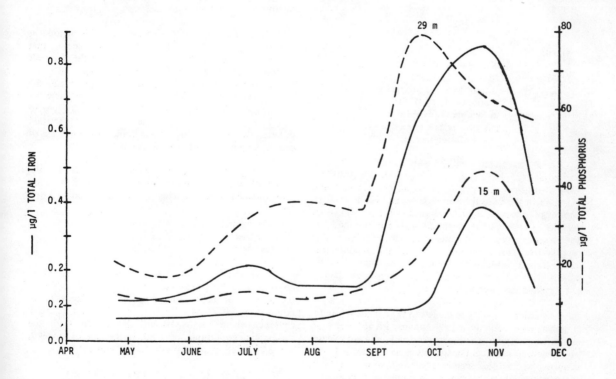

Figure 5. Relationship of total phosphorus and iron content near the bottom (29 m) and at 15 m during the stratified period in Lake Sammamish (Horton, 1972).

9

Model of Phosphorus Response

To determine if the lack of a significant change in phosphorus content in Lake Sammamish was predictable, gross compartment models were used to study the response of the lake total phosphorus concentration to step changes in the input of total phosphorus. Phosphorus input, outflow, and exchange with the sediments were included in the models. The lake is assumed to be two layered during the stratified period and uniformly mixed during the remainder of the year, and the water column P content is considered influenced only by the above mentioned gains and losses; complex cycling within the water column is ignored.

Model A is after Vollenweider (1969) who assumed that in addition to the input to and output from the lake there is a net loss of phosphorus to the sediments. He also assumed that this net loss is proportional to the concentration of phosphorus in the lake. Model B was modified from Vollenweider's by Lorenzen (no date) to the extent that the sediment and release rates were considered separately. The two models are:

$$(A) \quad \frac{dP}{dt} = \frac{M}{A} - 0.75 \frac{Q_o P}{ZA} - (S - R)$$

$$(B) \quad \frac{dP}{dt} = \frac{M}{A} - 0.75 \frac{Q_o P}{ZA} + R - S$$

in which

P	=	total amount of P in column (g/m^2)
M	=	input loading (g/yr)
Q_o	=	output flow rate (m^3/yr)
Z	=	depth of column (m^3)
A	=	surface area of lake (m^2)
R	=	sediment release rate ($g/m^2 \cdot yr$)
S	=	sedimentation rate ($g/m^2 \cdot yr$)

A P stratification adjustment factor of 0.75 was used in place of the effective flushing volume for stratified lakes where the outflow is from the surface waters. What is desired in the output term of the equation is the actual amount of P leaving the lake. Since the P is drained from the lake at epilimnion concentrations and in the model P is represented by a mean annual mixed water column, the appropriate factor is the ratio of mean annual epilimnetic concentration to the mean annual water column concentration. This ratio is the stratification adjustment factor and was 0.75 for Lake Sammamish.

In order to use these models to predict the response of phosphorus content in Lake Sammamish to diversion, values for phosphorus sedimentation and sediment release are needed. The annual phosphorus cycle was divided into anaerobic and aerobic segments according to when total P in the water column reaches a minimum and maximum—about May 31 and November 30, respectively. From the gain and loss from the water column during these periods along with the input and output of P, the net rates of release and sedimentation were estimated (Vollenweider, 1969). Five years of data were available.

The large pre-diversion maximums in November were attributed to extra strong mixing phenomena. If those are ignored then the water column maximum becomes 1.8 g/m^2, the minimum is 0.4 g/m^2 and the water column gains 1.4 g/m^2. The detention was 0.2 g/m^2 for the anaerobic period and 0.3 g/m^2 for the aerobic period. To obtain gross rates from these net rates a measured (*in situ* columns) sediment release rate of 0.5 $\mu g \cdot P/cm^2 \cdot$ day was used (Monahan, personal communication). This value becomes 0.9 g/m^2 for the 180 day anaerobic period.

To estimate gross annual rates for sedimentation and release, the following equation was used;

$$\Delta P \text{ content} = \text{Retention} + \text{Release} - \text{Sedimentation}$$

During the aerobic period release is assumed zero and an estimate of sedimentation is straightforward. However, during the anaerobic period true release is also unknown, because the measured values are from quiescent plastic columns. Since 1.4 g/m^2 > 0.9 g/m^2—Sedimentation + 0.2 g/m^2, additional release must have resulted from mixing in the water column.

Sedimentation rate during the anaerobic period was estimated at 1.0 g/m^2. This was done by considering that water column concentration was a constant the first half of the anaerobic period (May 31 to August 30) and therefore one half the release rate of 0.9 g/m^2 and the retention of 0.2 g/m^2 is equal to sedimentation. The value of 1.0 g/m^2 was used rather than 1.1 g/m^2 (2 x 0.55 g/m^2) because it was felt sedimentation rate would decrease in the last half of the anaerobic period because of decreasing phytoplankton productivity.

Using 1.0 g/m^2 as the sedimentation rate for the six-month anaerobic period, values for the annual steady state variables were estimated by the above equation and are:

Sedimentation	= 2.7 g/m^2
Release	= 2.2 g/m^2
Retention	= 0.5 g/m^2

The transient change in lake water P content in response to a 30 percent step reduction in P income is predicted by models A and B in Figure 6. The sedimentation rate was allowed to vary as a function of the P content, where the coefficient determined from the steady state conditions was used. In the case of model A net sedimentation rate was related to P content by,

$$(S - R) = KP; \text{ where K is the coefficient.}$$

However, in model B the release rate was assumed constant leaving the gross sedimentation rate to vary with

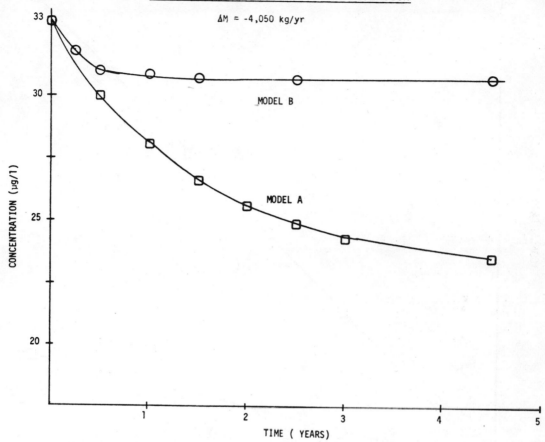

Figure 6. Transient change in total phosphorus concentration in Lake Sammamish predicted by two models in response to a 30 percent (4,050 kg/yr) decrease in phosphorus income.

P content according to,

$$S = K_1 P$$

These assumptions are the principal differences in the models that account for the differences in predicted response.

Model A predicts a sizable reduction in lake concentration of P while model B predicts a rapid, but only slight change (7 percent). The time to attain 98 percent of the change is 5.9 years for model A and 1.4 years for model B. The change predicted by model B is probably too small to have been detected, therefore this model fits reasonably well the observed lack of response.

The important assumption in model B is that sediment release rate is constant subsequent to diversion, and the rate is large because of the anaerobic nature of the lake. Thus, control of P content by sediment exchange in spite of moderate changes in P income is supported by these predictions. Response in Lake Washington was adequately predicted by model A (Emery et al., 1973).

Figure 7 shows further evidence of this. For the same response times the extent of change in P content in response to increases in P input was predicted to be much smaller with model B than A. Thus, assuming a constant sediment release rate a tripling of the P input rate to the lake over the past decades would have increased P content only 5 μ g/l according to model B.

11

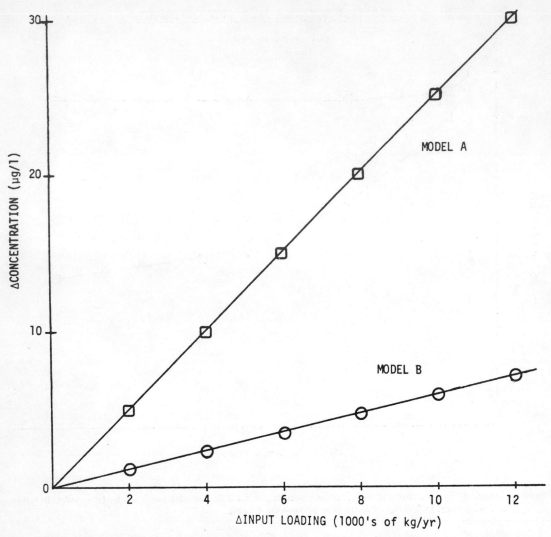

Figure 7. The change in steady state concentrations of total phosphorus in Lake Sammamish predicted by two models in response to changes in phosphorus income.

Summary

1. Lake Sammamish has not shown a significant response in five years following a diversion of about one third of its phosphorus income while nearby Lake Washington promptly responded with reduced photoplankton biomass in proportion to the decrease in P income and winter P concentration.

2. The probable explanation for this lack of response is that the P income change to Sammamish has occurred on the asymptote portion of a response curve that allows for much less lake response (Δ phytoplankton biomass) for a given increase in P income than was observed in Lake Washington. The internal mechanism that probably controls the P content available for vernal phytoplankton growth is large concentrations of iron released from sediments into an anaerobic hypolimnion, which complexes and subsequently resediments P before it is available to phytoplankton.

3. Strong control of water column concentration of total P by large exchanges of P with sediments was supported by a gross model predicting P response to income change. If sediment release rate was held constant and sedimentation rate was allowed to vary with P concentration then only a slight (7 percent)

reduction in concentration was predicted which would have occurred rapidly (1.4 years). Such a change could probably not be detected in view of the annual variations in P income and is therefore consistent with the hypothesis of P concentration control by the sediments in the range of P income studied. It is hypothesized that at higher rates of P income the binding capacity of the internal iron mechanism would be overcome and an increased rate of response would occur. This point may be near the two to three fold increase in P income that the lake would now be receiving if diversion had not occurred.

References

Andersson, G., G. Cronberg, and C. Gelin. 1973. Planktonic changes following restoration of Lake Trummen, Sweden. Ambio, 2:44-73.

Bush, R.M., and E. B. Welch. 1972. Plankton associations and related factors in a hypereutrophic lake. Water, Air and Soil Pollution, 1:257-274.

Edmondson, W.T. 1969. Eutrophication in North America. *In* Eutrophication: Causes, Consequences and Correctives. National Academy of Sciences, Proceedings of a Symposium, pp. 124-149.

Edmondson, W.T. 1970. Phosphorus, nitrogen, and algae in Lake Washington after diversion of sewage. Science, 169:690-691.

Edmondson, W.T. 1972. Nutrients and phytoplankton in Lake Washington. Limnol. and Oceanog., Special Symposia, 1:172-188.

Edmondson, W.T., G.C. Anderson, and D.R. Peterson. 1956. Artificial eutrophication of Lake Washington. Limnol. and Oceanog., 1:47-57.

Emery, R.M., C.E. Moon, and E.B. Welch. 1973. Delayed recovery in a mesotrophic lake after nutrient diversion. Jour. Water Pollut. Cont., 45:913-925.

Horton, M.A. 1972. The role of the sediments in the phosphorus cycle of Lake Sammamish. Unpublished M.S. thesis, Univ. of Wash., 220 p.

Kemmerer, G., J. Bovard, and W. Boormen. 1924. Northwestern lakes of the U.S.: Biological and chemical studies with reference to possibilities in production of fish. Bull. U.S. Bur. Fish., 39:51-140.

Ketelle, M.J., and P.D. Uttormark. 1971. Problem lakes in the U.S. EPA Technical Report 16010 EHR, 282 p.

Lorenzen, M. No date. Predicting the effects of nutrient diversion on lake recovery. Unpub. M.S., 14 p.

Mathiesen, J. 1971. Summer maxima of algae and eutrophication. Mitt. Verh. Internat. Verein. Limnol., 19:161-181.

Shapiro, J., W.T. Edmondson, and D.E. Allison. 1971. Changes in the chemical composition of sediments of Lake Washington, 1958-1970. Limnol. and Oceanog., 16:437-452.

Vollenweider, R. A. 1968. Scientific fundamentals of the eutrophication of lakes and flowing waters, with particular reference to N and P as factors in eutrophication. Organ. Econ. Co-Op. and Dev., Com. for Res. Co-Op., Paris. 192 p.

Vollenweider, R.A. 1969. Moglichkeiten und Grenzen elemtarer modelle der Stoffbilanz von Seen. Arch. Hydrobiol., 66:1-36.

MODELING ALGAL GROWTH DYNAMICS IN SHAGAWA LAKE, MINNESOTA,

WITH COMMENTS CONCERNING PROJECTED RESTORATION OF THE LAKE

D. P. Larsen, H. T. Mercier, and K. W. Malueg [1]

Introduction

During the past several decades much concern has developed over the continued deterioration of our nation's lakes, often characterized by nuisance algal blooms. Considerable resources have been expended to analyze causes and cures for this deterioration. Phosphorus has often been implicated as the element whose excessive supply to lake systems promotes abundant algal growth. In many cases, the excessive supply of phosphorus can be attributed to point sources, thus offering potential for control which should allow the lake system to recover. The Environmental Protection Agency has for the past several years pursued such a project at Ely in northeastern Minnesota, where the eutrophic condition of Shagawa Lake directly resulted from the fertilizing influence of that city's wastewater discharging directly into the lake from its secondary treatment plant.

Shagawa Lake has probably been eutrophic for at least several decades with local citizens commenting on excessive algal growths for over 40 years. The trophic state of Shagawa Lake is somewhat anomalous for most lakes in northeastern Minnesota are oligotrophic, located in heavily forested and sparsely populated terrain, often visited only by outdoor enthusiasts. Since the dominant source of phosphorus was from the wastewater discharge, an ideal opportunity existed to test the hypothesis that wastewater treated for phosphorus removal and allowed to enter a lake system would no longer exert its previous fertilizing influence, thus allowing the affected system to recover. It also afforded an opportunity to test and further develop conceptual and mathematical models which describe the eutrophication process and project changes due to management decisions. Presented herein is a model which describes algal dynamics within Shagawa Lake. It is also suggested that current models simulating lake recovery may be inappropriate for the Shagawa Lake system.

[1] D. P. Larsen, H.T. Mercier, and K. W. Malueg are with the Pacific Northwest Environmental Research Laboratory, Environmental Protection Agency, Corvallis, Oregon.

General Features of Shagawa Lake

Morphology

Shagawa Lake (Figure 1) is a relatively small lake with a surface area of approximately 9.3 km^2, a mean depth of about 5.7 m, and a volume of about 53 x 10^6 m^3. About 26 percent of the volume occurs below the mean depth. There are three "deep" holes of about 13.7 m each, a gradually sloping bottom, and steep sides. A survey conducted in 1937 indicated that the lake bottom was 70 percent muck, 29 percent sand, and the remainder coarser soils.

There is one major tributary, Burntside River, which drains nearby oligotrophic Burntside Lake, and several minor tributaries. The direct drainage basin is approximately 110 km^2.

Hydraulic budget (Table 1)

Approximately 70 percent of the water budget can be accounted for by Burntside River, which has an average annual discharge of 58.4 x 10^6 m^3, a maximum of 20 x 10^6 m^3 per month during spring runoff and a minimum of 1 x 10^6 m^3 per month during periods of low flow. The several minor tributaries account for 20 percent of the water input; rainfall accounts for another 8 percent. Wastewater accounts for less than 2 percent of the flow. The hydraulic residence time is approximately one year. The U.S. Geological Survey has estimated that groundwater influence is negligible.

Phosphorus budget (Table 1)

For the past six years approximately 80 percent of the annual phosphorus budget was attributed to the secondarily treated wastewater effluent and its bypass, for a combined load of approximately 6600 kg/yr. This compares with approximately 750 kg/yr from Burntside River. Rainfall and the lesser tributaries accounted for the remainder of the phosphorus entering the lake. An average annual loss of 4800 kg/yr occurred through Shagawa River, the outlet of the lake. There has been no evident increase in mean concentration of phosphorus in the lake over a six year period, thus the remainder must accumu-

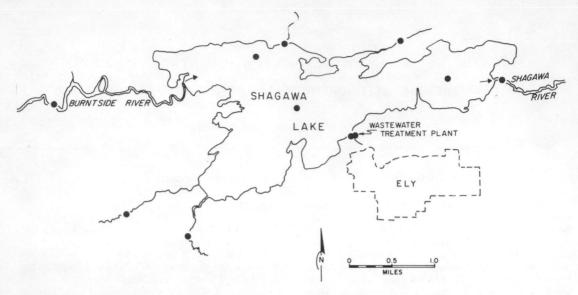

Figure 1. Map of Shagawa Lake.

late in the sediments and is approximately 30 percent of the annual total phosphorus input. The advanced waste-water treatment plant completed in the spring of 1973 removes in excess of 99 percent of the phosphorus entering it. This reduces the annual load to the lake by 80 percent.

Nitrogen budget (Table 1)

In contrast to the phosphorus budget, only about 23 percent of the annual input of nitrogen originated from the secondary wastewater system. Burntside River provided 45 percent of the nitrogen input, while rainfall and the lesser tributaries contributed about 32 percent. Nitrogen fixation has not yet been measured for Shagawa Lake; however, abundant growth of nitrogen-fixing blue green algae during periods of low ambient inorganic nitrogen indicate that nitrogen fixation may be an important source of nitrogen for the lake. Relatively little nitrogen removal occurs during advanced wastewater treatment, hence the nitrogen budget should remain essentially unaltered.

Algal growth dynamics

Since the major purpose of this presentation is to discuss a model which describes algal dynamics within Shagawa Lake, a brief summary of the pertinent features associated with algal growth is presented. The changes in surface concentration of algal biomass (as chlorophyll *a*), available phosphorus (as orthophosphate-phosphorus), and available nitrogen (as the sum of nitrate, nitrite, and ammonia) at one station for 1972 are summarized in

Figure 2. Except for relatively minor details, similar patterns are observed at other stations and other depths down to about 4.5 meters, the depth to which most pronounced algal activity occurs in this lake. The major features include a spring bloom occurring at approxi-mately the time of ice breakup when concentrations of

Table 1. Preliminary hydraulic, phosphorus and nitrogen budgets of Shagawa Lake—average for 1967-1972.

Tributary	Water m^3/yr x 10^6	Total Phosphorus kg/yr	Total Nitrogen kg/yr
INFLOW			
Burntside River	58.4	753	29,568
Longstorff Creek	7.3	132	5,467
Armstrong Creek	4.4	95	3,418
Burgo Creek	3.5	95	2,949
Bjorkman Creek	1.6	29	1,544
Precipitation-Evaporation	6.7	113	7,125
Wastewater	1.6	5,380	14,987
OUTFLOW			
Shagawa River	85.5	4,764	72,626
NET	- 2.0	+1,833	-7,568

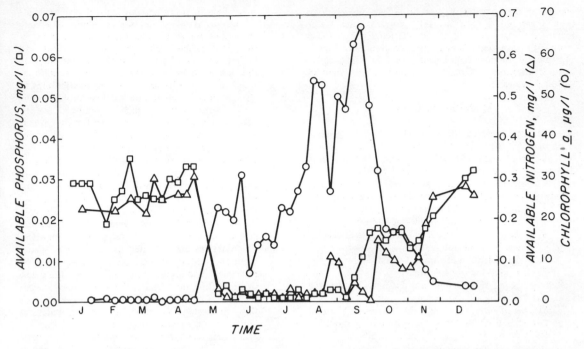

Figure 2. Surface changes in chlorophyll *a,* **available nitrogen, and available phosphorus in Shagawa Lake in 1972.**

both available nitrogen and available phosphorus decrease to limiting levels. Both nutrients are available in a ratio of about 10:1 and disappear at about that ratio. Shortly after the spring bloom, algal biomass declines presumably due to the low ambient concentrations of available nutrients as well as to their low rate of supply. In early July, although ambient levels of nutrients do not change appreciably, algal biomass increases rapidly reaching a maximum in late August of about 60 µg/liter of chlorophyll *a*. Only during late August do nutrient levels begin to increase, suggesting that the rate of supply of necessary nutrients has increased beyond the capacity of the algal community to consume them.

Fall circulation began during early September. During this period nutrients from the deep waters were transported into the photic zone, promoting a final burst of algal activity prior to the community die-off in later September.

A weekly mass balance analysis of total phosphorus within the lake during June, July, and August indicated that the increases in phosphorus concentration observed within the lake could not be accounted for by differences between inputs and outputs. To account for this difference, it is hypothesized that during this interval the sediments provided large amounts of phosphorus to the lake amounting to approximately 280 kg/week, while for the same period the sum of all measured sources accounted for 140 kg/week. The interval during which the

large release of phosphorus occurred correlates well with the period during which the deeper waters are anoxic. It is felt that this pulse of phosphorus promotes the large bloom of algae during July and August, and may provide a mechanism whereby recovery of the lake is delayed.

Modeling Concepts

Levins (1966) suggests that modeling strategies have developed along three lines, that of (1) sacrificing generality to gain realism and precision, (2) sacrificing realism to gain generality and precision, and (3) sacrificing precision to gain realism and generality. These strategies have developed in an attempt to reduce complex ecological systems into tractable models which assist in understanding, predicting, and managing natural systems.

Models of eutrophic systems tend to have been developed along lines 1 and 3. Vollenweider (1968, 1969) and Shannon and Brezonik (1972) have presented models which attempt to describe the trophic state of lakes based upon general features of lake systems—loadings, mean depths, hydraulic and nutrient retention times and the relationships between these features, and the level of productivity of the systems, hence sacrificing precision to gain generality and realism.

Others have developed models emphasizing mechanistic relationships between various elements of the systems and in so doing have tended to increase precision and realism at the expense of generality. Such models have been presented by various IBP programs, the Water Resources Engineers and the Manhattan College (New York), Environmental Engineering Program, and others. The present model development falls into this category since it attempts specifically to describe algal dynamics within Shagawa Lake.

Model Formulation

The model development follows the general scheme expressed by Kowal (1971) who states that model formulation of real world systems requires (1) operational definitions of the important real world variables, and (2) precise statements of the hypothetical relationships among these variables. The formulating process requires:
1. Specification of system of interest.
2. Specification of variables of interest.
3. Construction of control diagrams and classification of variables.
4. Specification of forms of equations.
5. Evaluation of constants.
For the purposes of modeling algal dynamics of Shagawa Lake, the following are considered.
1. *System of interest.* A simplified version of Shagawa Lake has been chosen as the system to be modeled. Its volume is 53×10^6 m^3, depth is 5.7 m throughout, and its contents are well mixed.
2. *Variable specification.* Since one of the basic purposes of modeling the eutrophication process is to understand the causes for changes in algal biomass, the variables chosen are (1) algal biomass, (2) factors which directly affect the ability of algae to

grow—light intensity and concentration of available nitrogen (nitrate, nitrite, and ammonia) and available phosphorus (orthophosphate), (3) rate of supply of available nitrogen and phosphorus, and (4) rate of flow of water. For the present purposes the model does not include such variables as zooplankton and bacteria and their relationships to changes in algal activity nor the relationship among various other variables such as temperature, pH, CO_2, and algal activity, assuming for the present that these variables exert a relatively small effect upon the general response of the system.

3. *Construction of control diagram and classification of variables.* Fairly complex diagrams can be constructed describing algal dynamics within lake systems; however, for the purpose of this model Figure 3 includes those variables which are considered important. Arrows connecting variables indicate the existence of relationships to be modeled. The variables are indicated as state variables and input and output variables. Input variables are "first principles" and are incorporated as measured fields, such as flow rates and nutrient loadings; output variables are functions of state variables and the flow of water. Variable units are specified as indicated in the control diagram.

4. *Specification of forms of equations.* The basic equation given for algal growth where growth is regulated by the metabolic properties of the algal population is:

$$\frac{dA}{dt} = K \cdot A$$

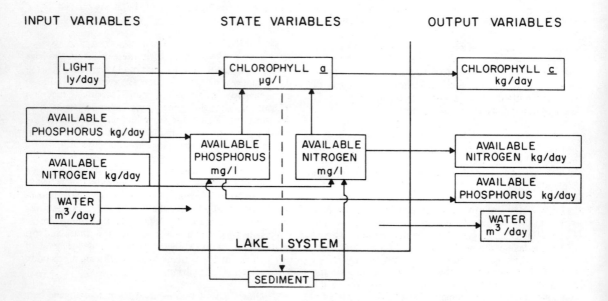

Figure 3. Control diagram of modeled lake system.

in which

A   =   algal concentration
K   =   growth coefficient
t   =   time

The growth coefficient, K, is taken as the maximum specific rate at which an algal population (community) can increase if all environmental variables are optimized for its growth, a situation rarely observed in natural systems. Growth is most often restricted by suboptimal light intensity and nutrient availability or their combination. Expressions have been developed to account for the restriction by these factors and the actual growth is often stated as a fraction of the maximum specific growth rate. Temperature affects growth, but its effect has not been included in the model.

Light limitation. For modeling algal dynamics within a well mixed lake, an expression was sought which would describe the fractional reduction in the maximum specific growth rate in the euphotic zone on a daily basis as it related to total daily radiation. That such an expression might exist was demonstrated by Ryther (1958) who related relative daily photosynthesis to total daily radiation for marine phytoplankton.

It was found that by averaging the expression (Steele, 1965; DiToro et al., 1971)

$$F\left[I(z,t)\right] = \frac{I(z,t)}{I_s} \cdot e^{\left[-\frac{I(z,t)}{I_s} + 1\right]}$$

in which

I   =   incident radiation
I_s =   radiation intensity at which maximum specific growth rate is achieved
z   =   depth
t   =   time
$F[I(z,t)]$ = relative photosynthesis

over the euphotic depth (taken to be the depth at which 1 percent of the surface radiation is obtained) and 24 hours an expression,

$$\frac{\overline{F}max \cdot R(t)}{K_r + R(t)}$$

in which

$\overline{F}max$ = maximum fractional reduction in daily specific growth rate over euphotic depth
R(t)   =   total daily radiation
K_r =   total daily radiation at which ½ of $\overline{F}max$ is attained,

was obtained to express the fractional reduction in K due to light limitation. The expression is reasonably valid as demonstrated by a fit to Ryther's (1958) data for $\overline{F}max = 0.35$ and $K_r = 163$ langleys/day. The data presented by Ryther (Figure

4) were divided by 24 hours and the product of the extinction coefficient and euphotic depth; if the euphotic depth is taken as the 1 percent light level, then the product of the euphotic depth and the extinction coefficient is 4.6.

Nutrient limitation. Several studies (Dugdale, 1967; Eppley and Coatsworth, 1968; Eppley et al., 1969; Eppley and Thomas, 1969; Fuhs, 1969; Fuhs et al., 1972) have indicated that a relationship between the specific rate of nutrient uptake by phytoplankton and nutrient concentration can be approximated by the following expression:

$$\mu = \frac{\mu_{max} \cdot N}{K_N + N}.$$

in which

μ =   specific rate of uptake
μ_{max} = maximum specific rate of uptake
N   =   nutrient concentration
K_N =   nutrient concentration at which $\mu = \frac{1}{2}\,\mu_{max}$

Hence the fractional reduction in μ_{max} is $N/(K_N + N)$. Although a realistic expression for growth is considerably more complex, this relationship has often been adopted to describe the nutrient dependency of growth. The maximum specific uptake rate then becomes the maximum specific growth rate, K.

Algal growth equation. Adopting a mass balance approach and assuming a constant volume and a well mixed system, the following growth equation can be written:

$$\frac{dA}{dt} = \left[\text{rate in - rate out + growth - loss}\right]\frac{1}{V}. \quad (1)$$

in which

V   =   lake volume, m^3
A   =   algal concentration, mg/l chlorophyll *a*
t   =   time, days

The supply of algae from tributaries is negligible, therefore, the rate in was assumed to be zero; rate out is given by $W(t) \cdot A$ where W(t) = rate of outflow of water, m^3/day.

The growth rate can be limited by either light, phosphorus, or nitrogen nutrients or their combination. A reasonable fractional reduction in growth rate can be calculated by taking the minimum of the expressions for light and nutrients at any time. Thus,

$$C_{lim} = \min\left[\frac{\overline{F}max \cdot R(t)}{K_r + R(t)},\; \frac{OP}{K_{op} + OP},\; \frac{AN}{K_{an} + AN}\right]$$

in which

$\overline{F}max$, R(t), and K_r have been defined previously

19

C_{lim} = the minimum fractional reduction in the growth coefficient K, due to light, available nitrogen, or available phosphorus

OP = available phosphorus concentration

K_{op} = available phosphorus concentration at which ½ maximum specific growth rate is achieved

AN = available nitrogen concentration (sum of nitrate, nitrite, and ammonia)

K_{an} = available nitrogen concentration at which ½ maximum specific growth rate is achieved.

Since algal growth only occurs over the euphotic depth, it was necessary to incorporate its value into the model. For modeling purposes it is generally assumed that no appreciable algal growth occurs beneath a depth at which 1 percent of surface illumination exists. Since light decays exponentially with depth, a relationship between extinction coefficient and euphotic depth can be obtained, thus

$$I = I_o e^{-\epsilon z}$$

in which

I_o = surface radiation
I = radiation at depth
ϵ = extinction coefficient.

Solving for z when $I = 0.01 I_o$, one obtains:

$$\text{euphotic depth} = \frac{4.6}{\epsilon}$$

The extinction coefficient is a function of the quality of the water and particularly of the amount of suspended algal material. Often a linear relationship is obtained between ϵ and algal material, thus $\epsilon = M \cdot A + B$ where M and B are regression constants.

The euphotic depth then becomes a function of algal concentration:

$$\text{euphotic depth} = \frac{4.6}{M \cdot A + B}$$

The growth term of Equation 1 is then $C_{lim} \cdot K \cdot A$ (euphotic volume). Since growth only occurs in a fraction of the lake and since in the idealized lake the ratio of euphotic volume to lake volume equals the ratio of euphotic depth to lake

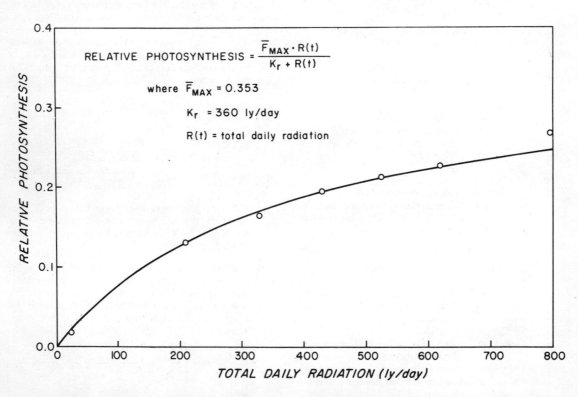

$$\text{RELATIVE PHOTOSYNTHESIS} = \frac{\overline{F}_{MAX} \cdot R(t)}{K_r + R(t)}$$

where $\overline{F}_{MAX} = 0.353$

$K_r = 360$ ly/day

$R(t) =$ total daily radiation

Figure 4. Relationship between daily relative photosynthesis over euphotic depth and total daily radiation after Ryther (1958).

depth, the specific rate of growth in the euphotic zone is reduced by the ratio of the euphotic depth to the lake depth to obtain a specific rate of growth within the entire lake.

Setting

$$C_g = C_{lim} \cdot \frac{euphotic\ depth}{lake\ depth},$$

$\frac{growth}{V}$ (from Equation 1) becomes

$$C_g \cdot K \cdot A.$$

Algal biomass declines due to such factors as zooplankton grazing, metabolic processes, sinking, etc.; however, these effects have been lumped into a generalized loss coefficient, C_L. Thus, loss in Equation 1 becomes $C_L \cdot A \cdot V$. The complete expression for the algal change in concentration within the lake then is:

$$\frac{dA}{dt} = (C_g \cdot K - C_L)A - Q(t) \cdot A \quad \ldots \quad (2)$$

in which

$$Q(t) = \frac{W(t)}{V}$$

Nutrient equation. The mass balance equation for nutrients is similarly expressed:

$$\frac{dN}{dt} = [rate\ in - rate\ out - rate\ consumed] \cdot \frac{1}{V} \quad (3)$$

Since algal biomass is expressed as chlorophyll *a*, it is necessary to convert chlorophyll *a* values to equivalent amounts of nitrogen and phosphorus. Often a constant yield coefficient is used, thus Equation 3 becomes:

$$\frac{dOP}{dt} = \frac{rate\ in}{V} - P \cdot C_g \cdot K \cdot A - Q(t) \cdot OP \quad (4)$$

for phosphorus, where P = phosphorus yield coefficient and

$$\frac{dAN}{dt} = \frac{rate\ in}{V} - N \cdot C_g \cdot K \cdot A - Q(t) \cdot AN \quad (5)$$

for nitrogen, where N = nitrogen yield coefficient.

Thus algal dynamics and nutrient concentrations are expressed as three coupled differential equations (Equations 2, 4, and 5). Input variables (rates of supply of nutrients, rate of flow of water, and total daily radiation) were entered in tabular form.

5. *Evaluation of constants.* Parameters as often as possible have been estimated from data collected on

Shagawa Lake and are summarized in Table 2. For the cases where parameter estimates were unavailable, values were obtained from the literature and are so indicated.

Simulation Runs

Since most of the parameters were estimated for the Shagawa Lake system, the first simulation runs were designed to test the model response against lake data and to estimate the sensitivity of the model response to variations in the loss coefficient, C_L, since a value or range of values was not estimated.

The following values were taken as initial conditions for all runs: OP = 0.022 mg/l; AN = 0.22 mg/l; chlorophyll *a* = 0.001 mg/l. Since the lake was ice-covered for nearly six months, values of total daily radiation were reduced by 90 percent for that interval (Wright, 1964). A fourth-order variable time step Runge-Kutta subroutine was used on the Oregon State University CDC 3600 computer to solve the differential equations.

Results

Simulations 1, 2, and 3

For the first three runs, C_L was chosen as 0.10 day^{-1} (Figure 5), 0.05 day^{-1} (Figure 6), and 0.025 day^{-1} (Figure 7). Based upon the initial conditions and the slightly different rates of supply of available nitrogen and phosphorus, a nitrogen limited system is produced. The spring bloom is simulated by the model, but its magnitude and timing are sensitive to the magnitude of the loss coefficient. As C_L decreases, the bloom magnitude increases and its peak occurs earlier. Further, after the initial bloom declines, the quasi-steady state levels of algae increase as the loss coefficient decreases.

The most remarkable difference between the simulated runs and the observed changes is that the large observed summer peak in algal concentration is not simulated. It is likely that other sources of nitrogen and phosphorus not included as tributary and rainfall produce this large bloom. Since the simulated runs demonstrate an adequate amount of available phosphorus to promote further growth, other sources of nitrogen must be hypothesized. Data for Shagawa Lake indicate that considerable amounts of phosphorus and nitrogen are released from the sediments during June, July, and August; nitrogen is released at approximately twice the rate that phosphorus is released. A second source of nitrogen not included in preliminary runs is atmospheric nitrogen dissolved in the water column. These sources of nutrients have been included in further runs as hypothesized inputs.

Simulations 4, 5, and 6

For simulation 4 (Figure 8), calculated values of nitrogen and phosphorus inputs from the sediments were used. Phosphorus loading was increased by 40 kg/day, and available nitrogen by 80 kg/day. The sediment discharge interval was from day 170 through day 230 (mid-June to mid-August). C_L was set at 0.05 day^{-1}.

A slightly greater algal biomass is produced by the additional nitrogen input than without the added input (compare Figure 8 to Figure 6). Phosphorus is supplied in excess of nitrogen relative to the algal demand, hence its magnitude increases markedly in the simulation run.

If it is assumed that the algal community can meet its nitrogen demand by nitrogen fixation and that its

Table 2. Estimates of model parameters.

	Parameter	Value	Source of Estimate
K	= maximum specific growth rate	2.4 day^{-1}	photosynthesis measurements
$\overline{F}$max	= maximum value for fractional reduction in specific growth rate	0.27	photosynthesis measurements
K_r	= total daily radiation at which one-half $\overline{F}$max is achieved	150 ly/day	photosynthesis measurements
I_s	= radiation intensity at which maximum photosynthesis is achieved	0.2 ly/min	photosynthesis measurements
K_{op}	= phosphate concentration at which one-half maximum specific growth rate is achieved	0.001 mg/liter	literature value
K_{an}	= available nitrogen concentration at which one-half maximum specific growth rate is achieved	0.014 mg/liter	literature value
P	= phosphorus-chlorophyll a yield coefficient	0.63 mgP/mg chl a	regression of particulate phosphorus on chlorophyll a
N	= nitrogen-chlorophyll a yield coefficient	7.2 mgN/mg chl a	regression of particulate nitrogen on chlorophyll a
M	= change in extinction coefficient due to changes in chlorophyll a	19 mg•L^{-1}•m^{-1}/mg•L^{-1}chl a	regression of extinction coefficient on chlorophyll a concentration
B	= extinction coefficient due to non-algal effects	0.16 m^{-1}	regression of extinction coefficient on chlorophyll a concentration

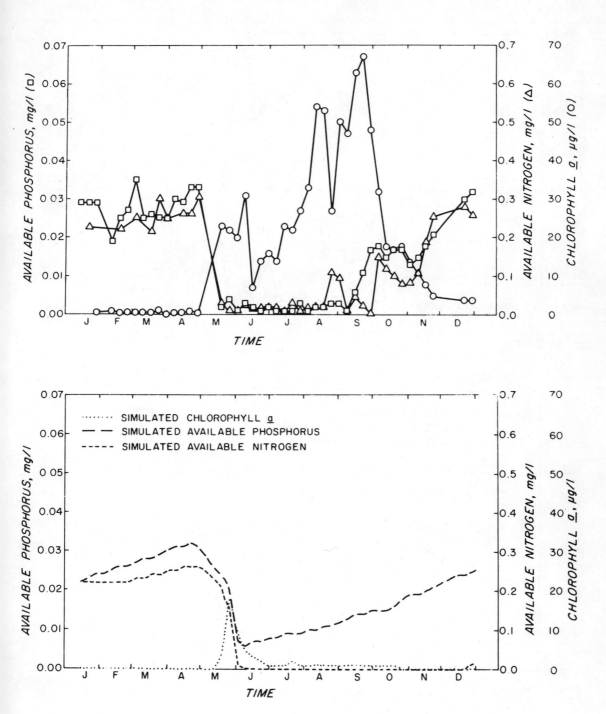

Figure 5. Simulation 1. $C_L = 0.10$ day $^{-1}$.

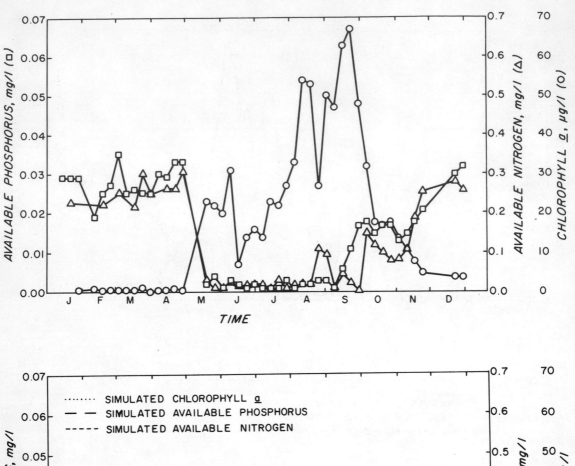

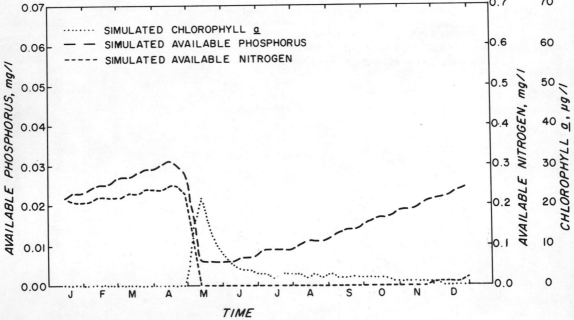

Figure 6. Simulation 2. $C_L = 0.05$ day^{-1}.

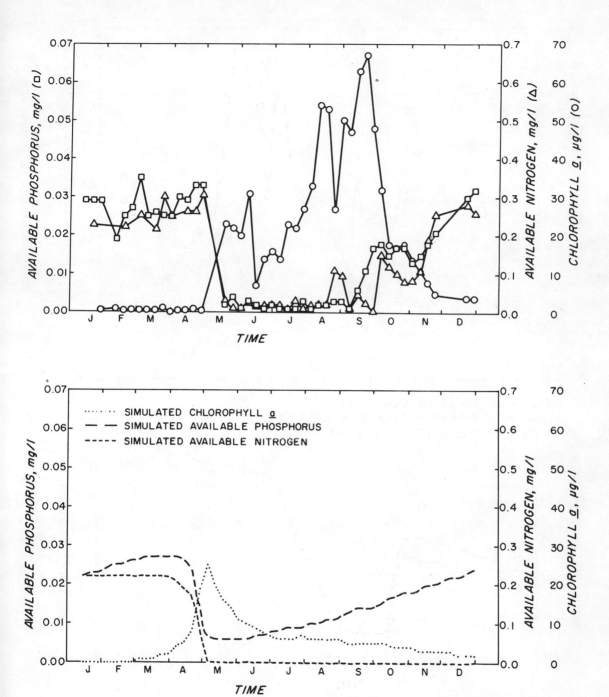

Figure 7. Simulation 3. $C_L = 0.025$ day^{-1}.

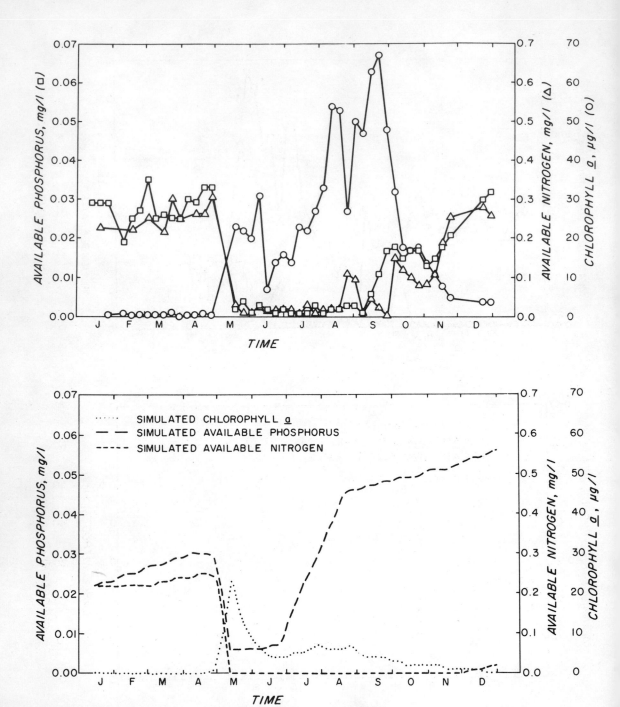

Figure 8. Simulation 4. C_L = 0.05 day^{-1}; added available phosphorus input = 40 kg/day; added available nitrogen input = 80 kg/day.

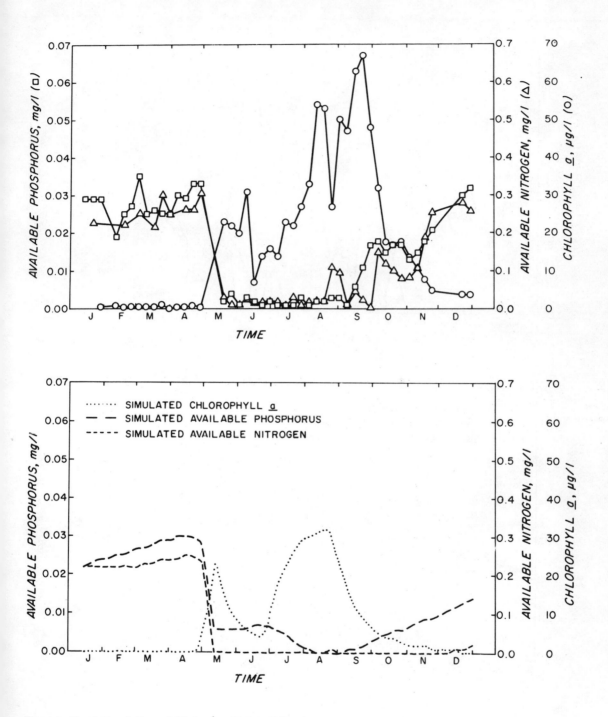

Figure 9. Simulation 5. C_L = 0.05 day^{-1}; added available phosphorus input = 40 kg/day; added available nitrogen input = 600 kg/day.

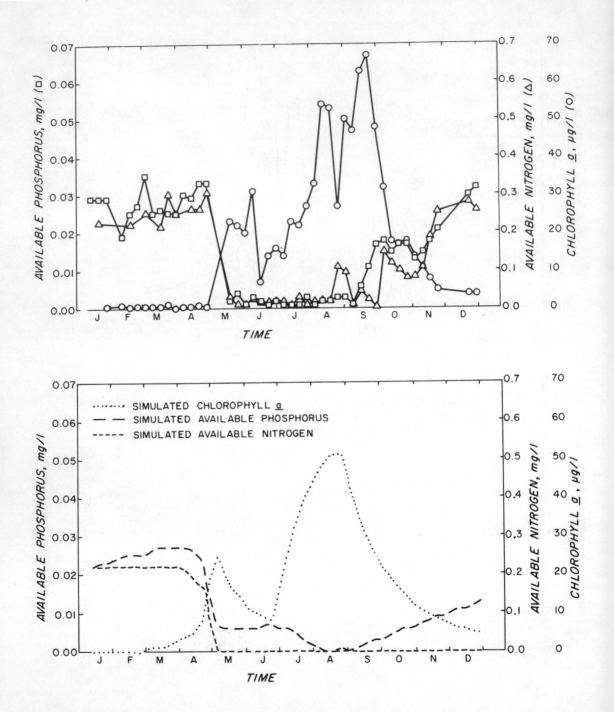

Figure 10. Simulation 6. C_L = 0.025 day^{-1}; added available phosphorus input = 40 kg/day; added available nitrogen input = 600 kg/day.

growth is regulated by the supply of phosphorus, a summer algal bloom can be simulated (simulations 5 and 6). For these two runs (Figures 9 and 10), tributary phosphorus loading (including wastewater and rainfall) was increased by 40 kg/day as above, and available nitrogen loading by 600 kg/day for the interval from day 170 through day 230. The loss coefficient was 0.05 day^{-1} (Figure 9) or 0.025 day^{-1} (Figure 10).

The simulated summer algal peak for run 6 is not quite as large as the observed value; further, its decline occurs considerably earlier than does the observed peak. When the simulated non-tributary sources of phosphorus and nitrogen are terminated, the algal biomass declines, whereas observed algal biomass does not decline for another month. The difference can be attributed to one of the model assumptions where a well mixed lake is considered. In actuality, about 80 percent of the lake is well mixed toward the end of summer; high concentrations of phosphorus and ammonia were observed in the deeper water. Fall circulation commenced in early September and apparently supplied sufficient nutrients for continued growth. The subsequent observed decrease in algal biomass perhaps occurred as a response to decreasing light and temperature fields; nutrient concentrations (phosphorus and nitrogen) were at non-limiting levels. These effects have not yet been included in the model formulation.

Discussion

The present model makes several simplifying assumptions concerning processes which are important in controlling nutrient and algal dynamics. The magnitude of these effects can only be assessed by observations within the particular system. For example, the net effect of zooplankton grazing has been incorporated into the loss coefficient as have the effects of sinking and metabolic loss. These effects produce variability in the loss of phytoplankton, and their magnitude may vary seasonally depending upon the composition of the algal crop and abundance of zooplankton. Thus, a diatom spring bloom may have a higher loss rate due to sinking and preferential grazing than would a blue-green summer bloom capable of flotation and undesirable to grazers. Further, sinking rates and metabolic losses are functions of the physiological state of an algal community which is in turn a function of the supply and demand of nutrients.

Processes which recycle nutrients within the water column have also been ignored, as has the physical chemistry of the inorganic nutrients.

Although a number of these simplifying assumptions have been made, the simulation model confirms the importance of two phenomena known to control the dynamics of algal populations. The most important is perhaps the supply of inorganic nutrients to the system. Vollenweider (1968, 1969) and Shannon and Brezonik

(1972) have demonstrated that the average level of productivity in lakes can be related to the rate of supply of nitrogen and phosphorus normalized to the lake surface area. For Shagawa Lake, sensitivity to supply rates is demonstrated by the lack of appreciable late summer response unless supply rates are increased by nitrogen fixation and hypothesized sediment supply of nutrients.

Secondly, the model demonstrates that the magnitude and timing of algal biomass levels are sensitive to the relative differences in the growth and loss terms. By changing the net loss term from 0.025 day^{-1} to 0.05 day^{-1}, a simulated summer algal peak decreases from 50 μg chlorophyll a/liter to slightly more than 30 μg chlorophyll a/liter. The values of the loss terms are relatively small and suggest that the net effect of several of the above processes which have been lumped into the loss term, though low in magnitude, can display remarkable control over algal abundances.

Predicted Restoration of Shagawa Lake

The full-scale restoration of eutrophic lakes requires considerable financial support; hence, few programs exist in which large scale manipulations are undertaken. Of those lakes which have received attention, perhaps the most notable is Lake Washington (Edmondson, 1972) where a reversal of the high level of algal activity has been documented after diversion of wastewater from the lake. Other systems receiving similar treatment have not responded as expected. For example, Lake Sammamish has shown little recovery five years after diversion (Emery et al., 1973). It is believed that the sediments play a dominant role in supplying nutrients to continue the high level of productivity in that lake.

Several investigators have presented models with which one might project the time course of parameters, such as total phosphorus, which reflect the level of productivity within lakes (Vollenweider, 1969; Megard, 1970; Sonzogni and Lee, 1972). These models are similarly structured and suggest that the rate of change of total phosphorus within a lake is the result of supply of phosphorus to the lake and losses from the lake through tributaries and loss to the lake bottom. The rate of loss to the lake bottom is generally assumed proportional to the amount present in the lake. By combining the effects of washout and sedimentation, a coefficient comparable to the hydraulic retention time—the phosphorus retention time—is obtained and is usually calculated as the quotient of the mean content of phosphorus and the annual loading of phosphorus, so that an equation relating change of phosphorus with time to gains and losses can be written as:

$$\frac{dP_t}{dt} = S - p \cdot P_t$$

in which

P_t = phosphorus content at time t
S = annual supply of phosphorus
p = reciprocal of phosphorus retention time

Models of this nature generally assume constant rates of inputs and losses based upon annual means and generally assume well mixed conditions, although the effect of a two layered system can be incorporated (Vollenweider, 1968).

If a lake originally in a state of equilibrium with respect to supply and loss of phosphorus is perturbed such that the supply is reduced, a new steady state will be achieved at a content the value of which depends upon the washout and sedimentation coefficients. If both are assumed constant and similar to their preperturbed values, then a new steady state level will nearly be attained in three to five times the phosphorus retention time.

For Shagawa Lake, the phosphorus retention time is approximately 0.5/year; the reduced annual loading is about 1300 kg/yr. Hence a new steady state level should be achieved in 1.5 to 2.5 years at which time the total phosphorus concentration should be nearly 0.01 mg/liter.

Shagawa Lake, however, seems to fall into that category of lake whose sediments will supply considerable amounts of nutrients to delay the projected recovery. The capacity of sediments to produce nutrients, particularly phosphorus, is dependent upon several complex factors (Lee, 1970) the most important of which is whether the overlying waters become anoxic. Thus, it becomes important to determine the extent to which anoxic conditions will continue to occur. This cannot yet be predicted. If anoxic conditions do not appear, it is likely that recovery of the lake will be quite rapid as might be predicted from the above model. If, however, the oxygen demand continues to be high, it is likely that the recovery will be prolonged, as in the case of Lake Sammamish.

Conclusions

A mathematical model relating algal growth dynamics to concentration of available nitrogen, available phosphorus, and total daily radiation within a well-mixed system has been presented. The model simulates the observed spring algal bloom and demonstrates sensitivity to the magnitude of a hypothesized loss coefficient in both timing and magnitude of the algal bloom. The model does not simulate the large late summer bloom of algae observed in the lake unless sediment nutrient supply and nitrogen fixation are included.

The model simulates the general cycle of available nitrogen and available phosphorus concentrations; however, unless nitrogen fixation is included, considerably higher levels of available phosphorus are produced than are observed in the lake. If the algal demand for available nitrogen can be met and if the rate of growth is regulated by the rate of supply of available phosphorus, then the model will simulate the low levels of both available nitrogen and available phosphorus observed during the summer months.

The observed increase in concentration of available phosphorus during fall and winter is simulated, but the observed increase in concentration of available nitrogen is not. Since external supply of both nitrogen and phosphorus has been included in the model, it is likely that the observed increases occur because algal demand has declined due to light and temperature effects such that supply is increased due to regeneration and exceeds demand.

The role of the sediments in providing critical nutrients to overlying waters has been demonstrated and it is likely that the sediments will extend the recovery of the lake beyond the 1.5-2.5 years projected by presently advanced models using a constant phosphorus residence time.

References

DiToro, D.M., D.J. O'Connor, and R.V. Thomann. 1971. A dynamic model of the phytoplankton populations in the Sacramento-San Joaquin Delta. Advances in Chemistry Series No. 106. Nonequilibrium Systems in Natural Water Chemistry. American Chemical Society. p. 131-150.

Dugdale, R.C. 1967. Nutrient limitation in the sea: dynamics, identification, and significance. Limnol. Oceanogr., 12:685-695.

Edmondson, W.T. 1972. Nutrients and phytoplankton in Lake Washington. In: Proceedings of the Symposium on Nutrients and Eutrophication: The Limiting Nutrient Controversy. G.E. Likens (ed.), Special Symposia, 1:172-193.

Emery, R.M., C.E. Moon, and E.B. Welch. 1973. Delayed recovery of a mesotrophic lake after nutrient diversion. J. Water Pollution Control Fed., 45(5):913-925.

Eppley, R.W., and J.L. Coatsworth. 1968. Uptake of nitrate and nitrite by Ditylum brightwellii - kinetics and mechanisms. J. Phycol., 4:151-156.

Eppley, R.W., J.N. Rogers, and J.J. McCarthy. 1969. Half-saturation constants for uptake of nitrate and ammonium by marine phytoplankton. Limnol. Oceanogr., 14:912-920.

Eppley, R.W., and W.H. Thomas. 1969. Comparison of half-saturation "constants" for growth and nitrate uptake of marine phytoplankton. J. Phycol., 5:365-369.

Fuhs, G.W. 1969. Phosphorus content and rate of growth in the diatoms Cyclotella nana and Thalassiosira fluviatilis. J. Phycol., 5:312-321.

Fuhs, G.W., S.D. Demmerle, E. Canelli, and M. Shen. 1972. Characterization of phosphorus-limited plankton algae with reflections on the limiting nutrient concept. *In:* Proceedings of the Symposium on Nutrients and Eutrophication: The Limiting Nutrient Controversy. G.E. Likens (ed.), Special Symposia, 1:113-133.

Kowal, N.E. 1971. A rationale for modeling dynamic ecological systems. *In:* Systems Analysis and Simulation in Ecology. Vol. 1. B.C. Patten (ed.), Academic Press, p. 123-194.

Lee, G.F. 1970. Factors affecting the transfer of materials between water and sediments. University of Wisconsin Eutrophication Information Program. Literature Review No. 1. p. 50.

Levins, R. 1966. The strategy of model building in population biology. American Scientist, 54(4):421-431.

Megard, R.O. 1970. Lake Minnetonka: nutrients, nutrient abatement and the photosynthetic system of the phytoplankton. Interim Report. No. 7. Limnological Research Center, Univ. of Minnesota.

Ryther, J.H. 1958. Photosynthesis in the ocean as a function of light intensity. Limnol. Oceanogr., 1:61-70.

Shannon, E.E., and P. Brezonik. 1972. Relationships between lake trophic state and nitrogen and phosphorus loading rates. Env. Sci. and Tech., 6(8):719-725.

Sonzogni, W.C., and G.F. Lee. 1972. Effect of diversion of domestic waste waters on phosphorus content and eutrophication of Madison Lakes. Unpublished manuscript.

Steele, J.H. 1965. Notes on some theoretical problems in production ecology. *In:* Primary Production in Aquatic Environments. C.R. Goldman (ed.), Mem. Ital. Idrobiol., 18 Suppl., University of California, Berkeley, p. 383-398.

Vollenweider, R.A. 1968. The scientific basis of lake and stream eutrophication with particular reference to phosphorus and nitrogen as eutrophication factors. Technical Report to O.E.C.D., Paris DAS/CSI/68, 27:1-182 (mimeogr.)

Vollenweider, R.A. 1969. Möglichkeiten und Grenzen elementarer Modelle der Stoffbilanz von Seen. Arch. Hydrobiol., 66(1):1-36.

Wright, R.T. 1964. Dynamics of a phytoplankton community in an ice-covered lake. Limnol. Oceanogr., 9(2):163-168.

SIMULATION OF URBAN RUNOFF, NUTRIENT LOADING, AND BIOTIC RESPONSE OF A SHALLOW EUTROPHIC LAKE[1]

D. D. Huff, J. F. Koonce, W. R. Ivarson, P. R. Weiler,
E. H. Dettmann, and R. F. Harris[2]

Introduction

It has been possible to describe various aspects of environmental processes on an individual basis for some time. For example, relationships between biological oxygen demand and dissolved oxygen are well known. Until recently, however, the means to simulate linked abiotic and biotic processes within a complete watershed ecosystem on a detailed basis have been out of reach. With increased understanding resulting from integrated multi-disciplinary ecosystem studies, and as a result of the development of digital computers with sufficient core memory and speed, detailed whole ecosystem simulation studies have become feasible. A contributing factor has been the willingness of large teams of investigators to work across disciplinary boundaries in sharing ideas and information.

In conjunction with research conducted under the aegis of the Eastern Deciduous Forest Biome (EDFB) of the International Biological Program (IBP), an attempt has been made to synthesize research and modeling results to simulate the behavior of the linked terrestrial and aquatic components of the Lake Wingra, Wisconsin, basin ecosystem. The models used have resulted from research both within and outside of EDFB programs, and represent the combined efforts of a very large number of individuals. In formulating the simulation study, emphasis has been placed on the use of relationships describing physical and biological processes mechanistically rather than empirical regression equations, since the former approach ultimately will yield a better understanding of the factors governing

ecosystem behavior. Thus the study reflects the current knowledge and ability to simulate the functioning of a lake basin ecosystem. It is felt that through an orderly progression of process model synthesis, testing, and new synthesis, a solid model structure can be built and extended for simulating whole ecosystem response to natural and man-made perturbations. With increased knowledge, the ability to predict accurately ecosystem response to new conditions will improve. At present, however, the reported simulation results are based upon a first attempt to link terrestrial and aquatic ecosystem models, and must be considered in that context. However, it is also true that the individual components of the whole basin simulation model have been extensively tested, and perform well by themselves.

In presenting the study, results of an analysis of the lake basin water balance are given first, followed by estimated nutrient loading from precipitation, dryfall, runoff, and groundwater flow. Finally simulations of the response of the lake through the period from April 10 to September 15, 1970, are compared with observations.

The Lake Wingra basin

Lake Wingra is located within the city limits of Madison, Wisconsin, and receives runoff from adjacent residential and natural (arboretum) areas. Table 1 presents a description of the physical properties of the basin, and Figure 1 shows important hydrologic features of the drainage area. Figure 2 presents a hydrographic map of Lake Wingra. Climatological data for the basin are available at Truax Field, a first-order weather station located about 8 kilometers from Lake Wingra. Hydrologic data are collected by the United States Geological Survey through a cooperative agreement with the EDFB program, and include records of storm drain runoff, spring flow, groundwater and lake levels, lake discharge, evaporation, and precipitation totals. Details of the monitoring system are available on request.

Simulation of runoff to Lake Wingra

Runoff to Lake Wingra from both natural and residential portions of the watershed was simulated using

[1]Support for the studies reported here was supplied in part by the Eutrophication Program of the University of Wisconsin, funded by the Environmental Protection Agency, and in part by the Eastern Deciduous Forest Biome, US-IBP, funded by the National Science Foundation under Interagency Agreement AG-199, 40-193-69, with the Atomic Energy Commission—Oak Ridge National Laboratory. Contribution No. 114 from the Eastern Deciduous Forest Biome, US-IBP.

[2]D.D. Huff, J.F. Koonce, W.R. Ivarson, P.R. Weiler, E.H. Dettmann, and R.F. Harris are with the Lake Wingra Study, Institute for Environmental Studies, University of Wisconsin, Madison, Wisconsin.

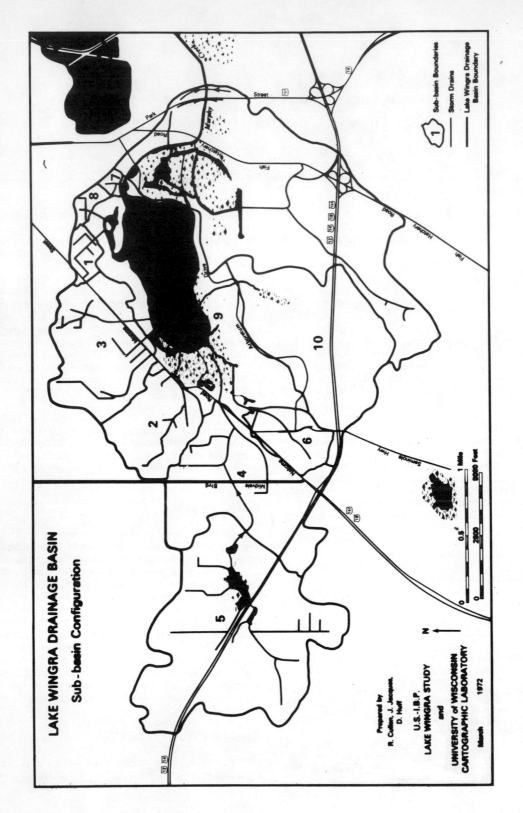

Figure 1. The Lake Wingra drainage basin.

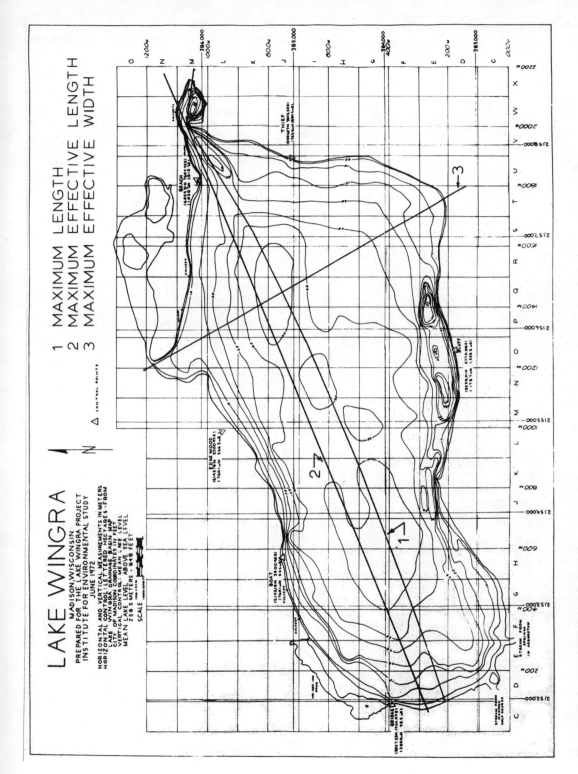

Figure 2. Hydrography of Lake Wingra.

Table 1. Physical properties of the Lake Wingra basin.[a]

Drainage Basin	
Land area draining to the lake	1437 Hectares
Lake surface area (including lagoons)	137 Ha
Total basin area	1574 Ha
Lake Hydrography	
Maximum Length	2.09 Km
Maximum Effective Length	2.16 Km
Maximum Width	1.11 Km
Mean Width	0.63 Km
Area	137 Ha
Volume (includes estimated lagoon volume)	3.35×10^6 m^3
Maximum Depth	6.10 m
Mean Depth	2.42 m
Length of Shoreline	5.91 Km
Shore Development Figure	1.45
Development of Volume	1.19

[a]Definitions of terms may be found in Hutchinson, G. E., "A Treatise on Limnology," John Wiley and Sons, Inc., New York, 1957.

the Hydrologic Transport Model (HTM) originally developed by Huff (1968) based upon the Stanford Watershed Model (Crawford and Linsley, 1966). Climatic data, including hourly precipitation, daily average wind speed, air and dew point temperatures, and solar radiation, were taken from Truax Field, Madison, Wisconsin, records. The basic time increment for model simulations was 15 minutes. The watershed was divided into several subbasins, representing the major storm drain outlets and the natural (arboretum) area. Runoff from all areas was simulated at 30-minute intervals, then summarized into daily and monthly rates.

For the purpose of representing each subbasin accurately, the subbasins were divided into two segments. One segment was composed of the impervious surface which contributes directly to the storm sewer system (i.e., streets, parking lots, and driveways). The other segment was composed of the pervious soil surface and those impervious areas tributary to the pervious surface (i.e., lawns, sidewalks, and roofs). Each segment was characterized by its own land, soil, and snow parameters. Measures of the fraction of subbasins represented by various forms of pervious and impervious surfaces have been presented by Cullen and Huff (1972). To introduce seasonal effects of infiltration rates (especially frozen soil), separate sets of winter and summer parameters were used. The variations between seasonal infiltration and interception parameters were determined primarily by trial and error. Detailed information regarding soil types and structure for the Lake Wingra basin has been developed by Huddleston (1972), and Huddleston, Luxmoore, and Hole (1972).

Figure 1 shows the channel system and subbasins for which calibration simulations were conducted. Since observed records of storm flows were not available for the period of interest, calibration studies used 1972 data.

Results of runoff simulations

The results of storm runoff simulations for water year 1972 have been compared directly with observed flows recorded in storm sewers (Nakoma and Manitou Way) and runoff from the arboretum area (Marshland Creek). These subbasins were chosen for model calibration because it was felt that they are representative of the full range of hydrologic responses to be expected in the Lake Wingra basin. Comparative graphs of calibration runs are provided (see Figures 3 to 9) for total runoff in inches per month for the water year 1972, mean daily flows for the month of December, 1971, and for an example storm hydrograph for December 15, 1971. Table 2 shows compared simulations and observations for calibration runs. The simulation results indicate that the model represents urban runoff events quite well. In fact, simulation results have been useful in detecting errors in observed records caused by ice effects during winter periods. Discrepancies between observed and simulated flows presented here demonstrate ice effect problems in recorded data. It is felt that during early spring, the simulation results are the more reliable flow values. Since results for ice free periods generally show that observed and simulated results agree within experimental error, it has been concluded that the runoff simulations for the period April-September 1970 are of accuracy comparable with that of later observations.

Table 2. Observed vs. simulated runoff volume for the Manitou Way and Nakoma storm sewer in inches per month for water year 1972.

	Manitou Way		Nakoma	
	Observed	Simulated	Observed	Simulated
Oct.	.11	.09	--	.08
Nov.	.35	.48	--	.44
Dec.	.38	.39	.44	.37
Jan.	.02	.07	.03	.06
Feb.[a]	.03	.08	1.10	.15
March[a]	.86	.66	1.94	1.11
April[a]	.55	.53	.49	.83
May	.32	.19	.28	.20
June	.08	0	.02	0
July[a]	.05	.24	.09	.27
Aug.	.93	.88	.94	.84
Sept.	.53	.45	.51	.43
Yearly Totals	4.21	4.06	6.36	4.70

[a]Month with possible error in observed data.

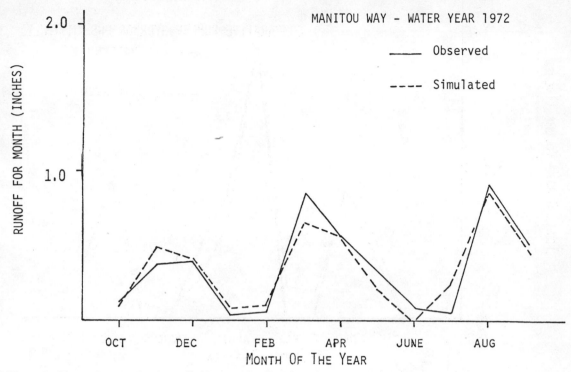

Figure 3. Observed vs. simulated runoff—Manitou Way (1972).

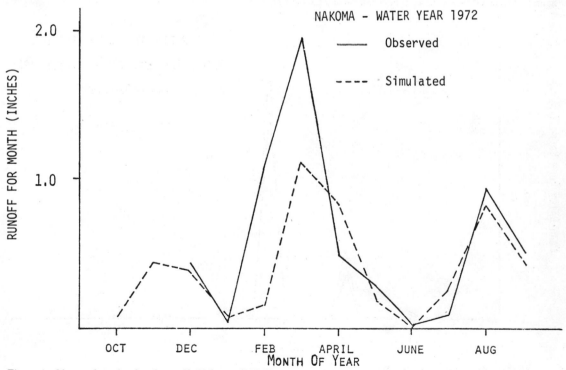

Figure 4. Observed vs. simulated runoff—Nakoma (1972).

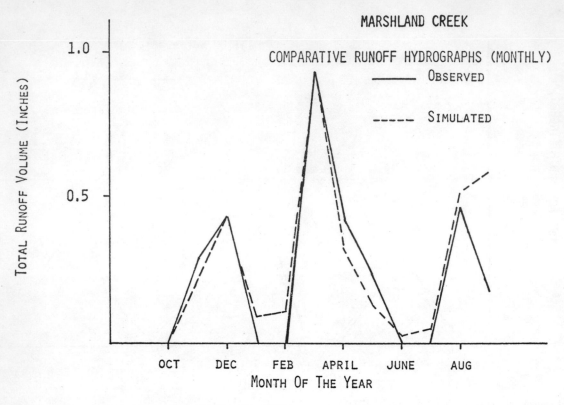

Figure 5. Observed vs. simulated runoff—Marshland Cr. (1972).

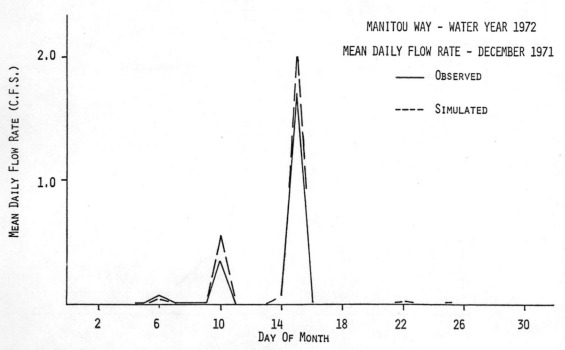

Figure 6. Mean daily flows—Manitou Way (December 1971).

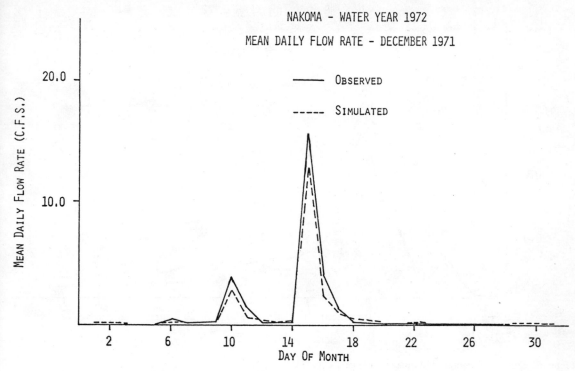

Figure 7. Mean daily flows—Nakoma (December 1971).

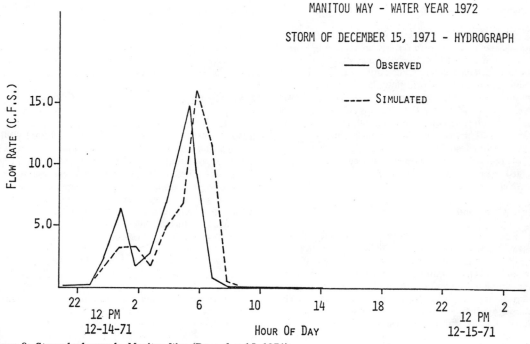

Figure 8. Storm hydrograph—Manitou Way (December 15, 1971).

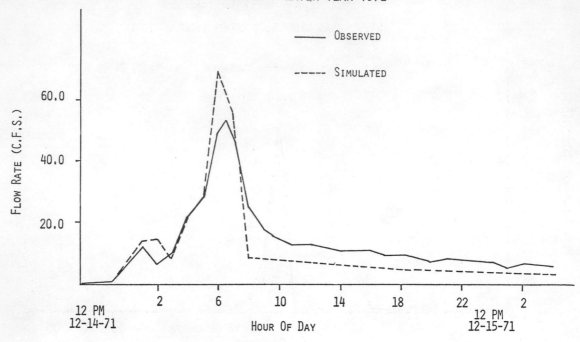

Figure 9. Storm hydrograph—Nakoma (December 15, 1971).

Estimation of groundwater flows

Groundwater flow to Lake Wingra comes both from major springs and extensive seepage areas. Springflow to the lake is monitored routinely at present, but observations were not available for the entire period of interest for simulation studies. Springflow data were lacking for April and May 1970, and were estimated. Values of 1972 simulated recharge for the three months antecedent to the month of interest were correlated with corresponding observations. Then the correlation relationships were used for 1970 estimates of springflow. Comparative results for one of the largest springs are presented in Figure 10. Ungaged seepage amounts were estimated from water balance computations.

Lake discharge and stage

Records for lake stage beginning in late April 1970 are available for Lake Wingra. The stage records, together with a theoretical rating curve for the lake outlet weir, were used to check water balance simulations. By comparing simulated and observed lake stage, it was possible to derive monthly values for groundwater seepage to the lake. As expected, seepage was highest in April, and declined steadily through September, as indicated in the water balance summary shown in Table 3. After adjusting for ungaged groundwater flow on a monthly basis,

simulated lake stage values were found to be within 0.01 feet of observations most of the time, and thus are generally within experimental error. Figure 11 shows the comparison between observed and simulated lake stage. For the aquatic ecosystem response studies, only simulated lake discharge values were used.

Lake evaporation

Lake evaporation was calculated on a daily basis using the method described by Roberts and Stall (1967), and climatic data from Truax Field. No corrections were made for advection or changes in heat storage in the lake for these studies. It is anticipated that such adjustments could be significant during the early part of the period as the lake warms to summer temperatures, but probably would not be important to the overall water balance computations for the full period of interest, since Lake Wingra has a relatively small heat storage capacity.

Results of whole basin water balance simulations

In general, results of the basin water balance simulations are quite satisfactory. Table 3 shows the quantitative water balance for the period of interest. To summarize, 25 percent of total inflow comes from storm drains, 30 percent from direct precipitation on the lake

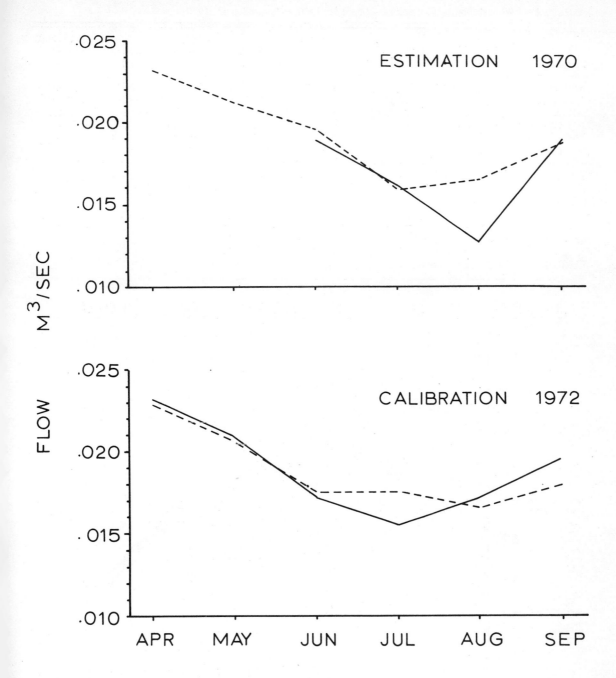

Figure 10. Estimation of arboretum spring flow (1970 and 1972).

Table 3. Lake Wingra water balance summary (April 10, 1970, to September 15, 1970).

Period	Precipitation (M^3)	Evaporation (M^3)	Groundwater Inflow (M^3)	Groundwater Loss (M^3)	Surface Inflow (M^3)	Lake Outflow (M^3)	Lake Storage Change (M^3)
APR 10-30	81,000	81,400	230,600	25,700	73,000	282,100	- 4,600
MAY 1-31	211,000	142,000	242,800	38,000	218,600	434,200	58,200
JUN 1-30	78,300	177,000	205,800	36,800	28,300	184,000	- 85,400
JUL 1-31	83,800	183,300	137,400	151,900	41,800	6,800	- 79,000
AUG 1-31	33,600	158,300	117,100	227,800	14,900	0	-220,500
SEP 1-15	189,100	46,100	78,600	169,000	194,900	24	247,500
TOTAL	676,800	788,100	1,012,300	649,200	571,500	907,100	- 83,800
INFLOW %	29.9	---	44.8	---	25.3	---	---
OUTFLOW %	---	33.6	---	27.7	---	38.7	---

PERIOD TOTAL INFLOW = 2,260,600 m^3

PERIOD TOTAL LOSSES = 2,344,400 m^3

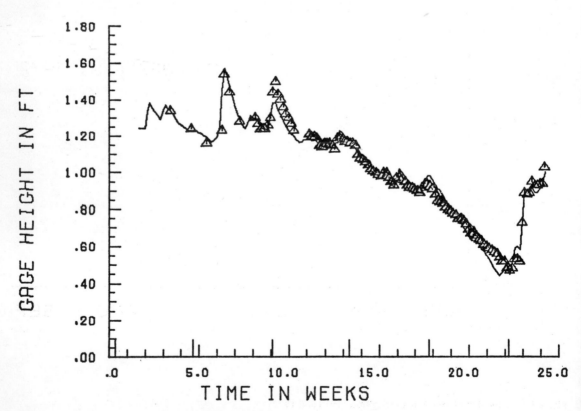

Figure 11. Observed and simulated lake stage (April 10 - September 15, 1970).

surface, and 45 percent enters as groundwater. Total outflow is divided among 33 percent as evaporation, 28 percent as groundwater loss, and 39 percent as outflow over the outlet spillway. Storage was 83,800 cubic meters lower at the end of the period than at the start. Groundwater loss has been estimated by difference from water balance calculations, hence will contain the net error from other water budget terms. However, the fractions of inflow quantities are in agreement with those given by Kluesener (1972) for 1970-71, and the other loss terms are well enough documented to give confidence in the estimated groundwater balance.

From the point of view of total surface runoff, the simulated monthly results appear to be as accurate as observations during periods of record, and thus provide valuable information in studying lake response for periods when flow observations are unavailable. The ability to reproduce lake stage using simulated flows has given further confidence in the overall accuracy of the flow estimates on a day to day basis. Since it has been concluded that inflow volume is extremely important to estimating nutrient loading (Harris et al., 1972), the hydrologic simulations were very important to the study.

Perhaps of additional interest is the potential to explore the relationship of runoff from impervious areas to total nutrient transport to the lake. Simulated runoff is differentiated into that from impervious and pervious areas, providing an opportunity to examine relationships between the fraction of runoff originating from a particular source area and the corresponding nutrient content of the runoff water. Data are available for studying this question, and investigations are currently underway.

Nutrient Sources

The nutrients simulated by the open lake model are dissolved phosphorus and dissolved mineral nitrogen. The latter includes both ammonium and nitrate nitrogen. The external sources of these nutrients are urban runoff from storm sewers draining into the lake or surrounding marshes, springwater, groundwater seepage, rainfall, and dryfall. Internal nutrient regeneration is also an important source of nutrients and will be considered in the section on open lake model simulations.

Most of the dissolved inorganic phosphorus entering the lake comes from urban runoff. According to the estimates made by coupling observed concentrations in runoff water to the hydrologic transport model, 84 percent of the dissolved inorganic phosphorus entering the lake system during the period under study (10 April 1970 to 15 September 1970) enters from the storm sewers. One storm sewer alone, Nakoma, accounts for 34 percent of the loading. This storm drain has generally lower concentrations of phosphorus than the other drains, but has a steadier flow of water. About two-thirds of the two square mile area from which Nakoma receives runoff

drains into a holding pond with a regulated spillway into the storm sewer system (Harris et al., 1972). In the model, this storm sewer is treated as a separate source distinct from the other storm sewers. Different sets of phosphorus concentration estimates are used for the Nakoma storm drain and other storm sewers (Harris et al., 1972). Other sources of dissolved inorganic phosphorus include: rainfall (5 percent), dryfall (5 percent), and springs and groundwater seepage (6 percent). Again estimates of the concentration in these sources are based on actual measurements (Kluesener, 1972).

Figure 12 shows the dissolved inorganic phosphorus input to Lake Wingra from 10 April to 15 September 1970. The peaks in this figure occur on days of sizable rainfall. The truncated peak in the seventh week reaches a height of 35 kilograms per day. Very little phosphorus enters the lake between storms.

Although springs and groundwater seepage provide little of the phosphorus entering the lake, they provide most of the dissolved mineral nitrogen. Sixty-five percent of the dissolved mineral nitrogen entering the lake during the period of interest came from this source. This nitrogen is almost entirely nitrate nitrogen. Estimates of spring and groundwater seepage nitrogen loading are based on measurements of surface spring concentrations (Kluesener, 1972). About 11 percent of the dissolved mineral nitrogen loading came from urban runoff, with rainfall and dryfall accounting for about 12 percent each (Kluesener, 1972). Figure 13 shows the daily dissolved mineral nitrogen input to the lake. As in the case of phosphorus, peaks of input occur during storms. Springs and seepage groundwater, however, provide significant amounts of nitrogen to the lake between storms.

Coupling between hydrologic and open lake models

Coupling between the hydrologic transport model and the open lake model was accomplished in the following manner. An HTM simulation of the basin was made for the period 10 April to 15 September 1970. Daily totals of water entering and leaving the lake from all major sources were computed. These values served as the basis for Table 3 presented earlier.

Losses of nutrients and plankton over the outlet dam were also computed daily from outflow data provided by the HTM simulation. The fractional rate of loss of nutrients and plankton to this outlet was assumed to equal the ratio of the daily outflow to the mean volume of the lake. This loss rate was always less than 1 percent per day. During much of the summer it was zero because no outflow occurred.

To obtain rates of nutrient loading from each source, the open lake model multiplies the amount of water entering from each source by the concentration of the nutrients in that source. Each day the mean input rates for both nutrients are calculated and used in the differential equations for the open lake nutrient concentrations. The nutrients are assumed to distribute themselves evenly throughout the lake. This assumption is justified for the open water zone by the shallow, well-mixed nature of the lake. A complication is the fact that most of the runoff must pass through a marsh or weedbed before entering the open water zone of the lake. The marsh and littoral zones may absorb or release nutrients and thereby alter the quantity of nutrients reaching the open lake. Such effects are difficult to study, and as yet there is not enough information to determine them in Lake Wingra. In order to compensate, the input rates may be multiplied by a constant factor. (See section on open lake model simulation.)

Open Lake Model

Simulation of the response of lake biota to nutrient flux is possible at several levels of resolution. Research in the Eastern Deciduous Forest Biome, has developed detailed models for many processes in aquatic ecosystems. Various combinations of these models have also been assimilated into ecosystem and subecosystem models (e.g., Park et al. and MacCormick et al., 1972). As a starting point, a subecosystem model (WINGRA 2) of the pelagic zone of Lake Wingra will be employed. Although this model has been specially applied to a shallow, temperate zone lake, the use of more general ecosystem models (e.g., CLEAN, Park et al.) is not precluded and will be one of the objectives of future work.

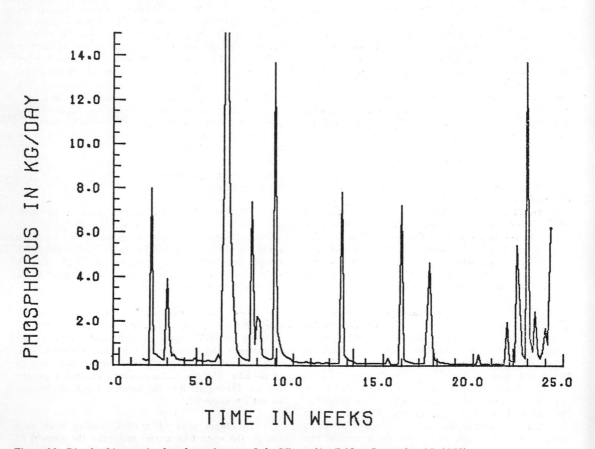

Figure 12. Dissolved inorganic phosphorus input to Lake Wingra (April 10 to September 15, 1970).

WINGRA 2 has already been extensively documented (MacCormick et al., 1972) and no attempt will be made here to discuss the details of the model. Instead, some of the general features of the model will be summarized. A flow diagram, Figure 14, and a transfer matrix, Figure 15, show the essential features of the model. The model is a nine compartment representation of the open water portion of a shallow lake. Each of the compartments can be subdivided into two or more subcompartments to more accurately resemble the range of adaptive strategies in multiple species communities. The individual compartment models are theoretically derived, mechanistic models. The whole open water model is numerically integrated over time with a fifth-order, variable step size Runge-Kutta algorithm and is programmed in Fortran V.

The model is driven by solar radiation, temperature, and nutrient concentration. Organic material inputs from the terrestrial and littoral ecosystems may also be accommodated. By incorporating the nutrients (nitrogen and phosphorus) as compartments in the model, nutrient concentration may also be dynamically simulated, and the hydrologic regime may thus be used to drive the model.

In its present form, the model requires several assumptions. Table 4 lists the major assumptions. These assumptions clearly limit the generality of the model and treat as constant some important physical, chemical, and biological processes. The incorporation of these assumptions can only be justified on the grounds of expediency. Their adoption has caused some problems in the adequacy of the simulations, and they explicitly constitute areas for further improvement. Nevertheless, the model has been instructive despite the limitation of these assumptions, and has proven to be a catalyst for better understanding of ecosystem functioning.

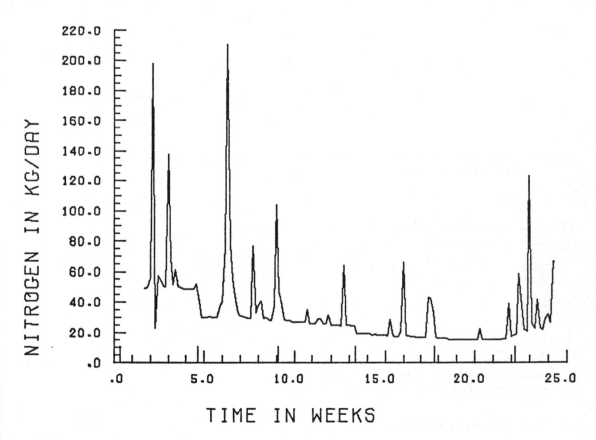

Figure 13. Dissolved mineral nitrogen input to Lake Wingra (April 10 to September 15, 1970).

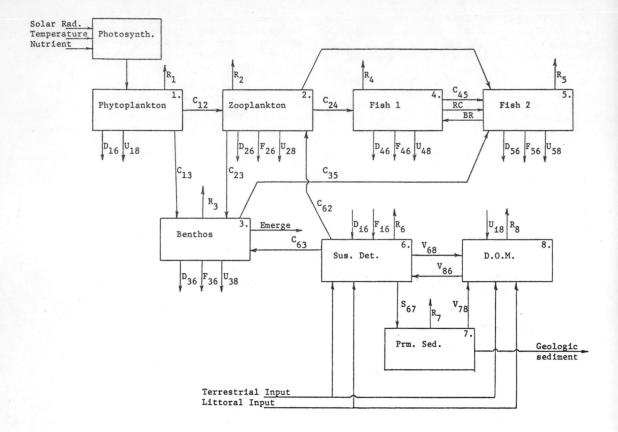

Figure 14. Flow diagram showing the pathways for transfer of biomass assumed in the open water model (after MacCormick et al., 1972).

Table 4. Assumptions of the open lake model.

1. The lake is uniformly mixed and turbulent mixing is constant throughout the year.

2. Marsh and littoral zones intercept a constant proportion of nutrients in runoff.

3. Only dissolved nutrients are transported to the open water zone of the lake.

4. Nutrient regeneration is independent of the biomass of organisms present in the open water zone of the lake.

5. Light intensity may be simulated as a sinusoidal function.

Model development

Any model reflects the system conceptualization of those who formulated it. These biological models are no exception. The process models that are the foundation of the open water model originated in working sessions between experimental, field, and mathematical ecologists. These models, therefore, are the best statements of the mechanisms regulating growth and reproduction of aquatic organisms. As such, they contain more detail than can be adequately expressed here. The writers would, therefore, like to give an overview of the rationale for model development.

One of the major areas of concern over eutrophication is the classical problem of algal bloom formation. The dynamics of algal biomass and photosynthesis for Lake Wingra phytoplankton are not unlike those for many other temperate lakes. Biomass and productivity of Lake

Wingra phytoplankton are shown for a fourteen month period in Figure 16. Model output, of course, is to be judged against this temporal pattern. Total algal biomass dynamics is, however, only a summary of a continuously changing assemblage. Many indices of this change may be found. Class composition, for example, shows a pronounced seasonal change that reflects the changing abundance of a few species (Figure 17). Growth form is also a dynamically changing characteristic of phytoplankton assemblages (Figures 18 and 19).

Seasonal succession is not limited to phytoplankton associations. Zooplankton and benthos also change seasonally, and succession over longer time intervals

BIOMASS FLOW MATRIX

INPUT (rates)	1	2	3	4	5	6	7	8	OUTPUT (rates) Respiration	Emergence	Geol. Sed.
Terrestrial						X		X			
Littoral						X		X			
Phytopl. Prod.	X										
BIOMASS POOLS											
Phytoplankton B_1		C_{12}	C_{13}			D_{16}		U_{18}	R_1		
Zooplankton B_2			C_{23}	C_{24}	C_{25}	D_{26} F_{26}		U_{28}	R_2		
Benthos B_3					C_{35}	D_{36} F_{36}		U_{38}	R_3	X	
Fish 1 B_4					C_{45} RC	D_{46} F_{46}		U_{48}	R_4		
Fish 2 B_5				BR		D_{56} F_{56}		U_{58}	R_5		
Sus. Det. B_6		C_{62}	C_{63}				S_{67}	V_{68}	R_6		
Prm. Sed. B_7								V_{78}	R_7		X
D.O.M. B_8						V_{86}			R_8		

Figure 15. Biomass flow matrix for open water model (after MacCormick et al., 1972).

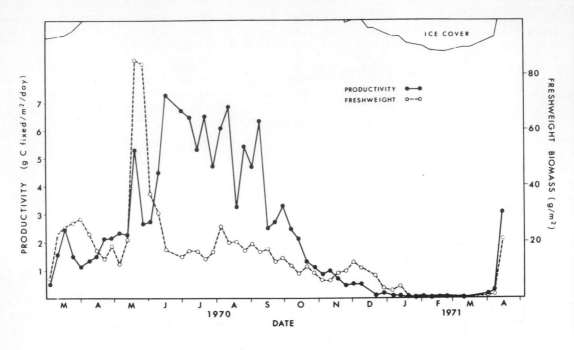

Figure 16. Average daily productivity and freshweight biomass of phytoplankton in Lake Wingra (after Koonce and Hasler, 1972).

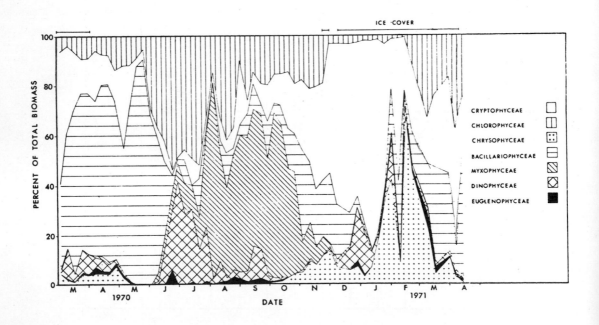

Figure 17. Class composition pattern for phytoplankton in Lake Wingra (after Koonce and Hasler, 1972).

48

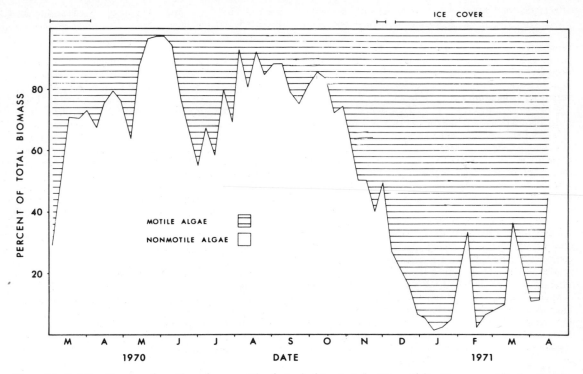

Figure 18. Relative biomass of motile and non-motile phytoplankton in Lake Wingra (after Koonce and Hasler, 1972).

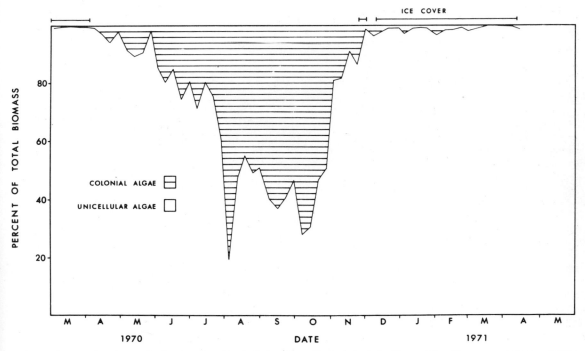

Figure 19. Relative biomass of colonial and unicellular phytoplankton in Lake Wingra (after Koonce and Hasler, 1972).

affects fish and other macro-components of an aquatic ecosystem. The challenge to a model, therefore, is to represent the basic mechanisms controlling these succession patterns. Without such a capability, the model is poorly suited for simulation outside the range of conditions for which it was derived.

Parameter estimation

Before describing the performance of this modeling effort, some attention should be given to the problem of obtaining parameter estimates for detailed models such as those employed in WINGRA 2. Because they are detailed statements of ecological processes, the models are characterized by rigorous parameter requirements. The phytoplankton equations are a case in point.

Phytoplankton growth is modeled as a function of the internal concentrations of carbon, nitrogen, and phosphorus. The internal pools of phosphorus and nitrogen, in turn, are related to external nutrient pools by a simple Michaelis-Menten type relationship. Similarly, the internal reduced carbon pool is maintained by carbon dioxide fixation through photosynthesis. The overall biomass dynamics is then a function of growth, zooplankton grazing, and sinking. The phytoplankton is, therefore, characterized by parameters relating to photosynthesis, respiration, nutrient uptake kinetics, growth rate, grazing

susceptibility, and rate of sinking. Needless to say, independent estimation of each of these parameters is a formidable task.

This parameter estimation problem is greatly facilitated if the process investigators are able to interrelate these parameters in a theoretical manner (e.g., Allen and Koonce, 1973, and Koonce and Hasler, 1972). Large phytoplankton species, for example, may have little grazing pressure, but by virtue of a small surface to volume ratio, have lower growth rates, lower nutrient affinities, and greater sinking rates than do small species subject to high grazing pressure. These theoretical relationships, however, require the participation of specialists in many disciplines and, therefore, almost require an interdisciplinary program. The parameter estimates here employed are, in fact, products of such interdisciplinary consultation.

Open lake model simulations

The open lake model being used in these simulations has gone through several improvement stages. The first simulations relied on observed nutrient concentrations, temperature, and light to drive the model. The objective was to assess the adequacy of this model and parameter estimates in reproducing observed dynamics of the lake biota. The first results, Figure 20, were from simulations

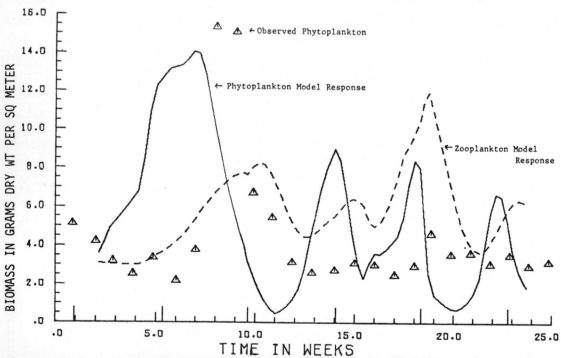

Figure 20. Open water model output from phytoplankton and zooplankton. Time varying constants are assumed in the model. Pelagic zone model output for phytoplankton and zooplankton. Solid line is simulated phytoplankton, dashed line is simulated zooplankton, and triangles represent observed phytoplankton data from 1970.

using only one species group in each compartment, but with time varying constants (MacCormick et al., 1972). The output is clearly not reproducing the observed patterns, but the model gives better simulations than that in which no time varying constants were used.

In this first attempt, the researchers tried to combine all of the adaptive strategies of a given trophic compartment into one general representation. Such a simplification was contradictory to the major species succession patterns observed in the lake. Therefore, the phytoplankton was subdivided into four strategically dissimilar groups. The results of these simulations are presented in Figures 21 and 22. The partitioning of the phytoplankton compartment improved the simulations, and a more realistic succession pattern evolved.

Using the parameters of the previous simulation, the model was then allowed to simulate water column nutrient concentrations, including the effects of hydrologic inputs and outputs, biological uptake, sedimentation losses and regeneration. Because the amount of nutrient retained by the marsh and littoral zones through which surface runoff must pass before reaching the open water of the lake presently cannot accurately be forecast, there are no direct measures of actual nutrient input to Lake Wingra. To circumvent this problem, the quantity of nitrogen and phosphorus reaching the open water from hydrologic sources was allowed to vary from 5 to 150 percent of that indicated by measurements and HTM simulations. Such a range, it was felt, might incorporate both efficient nutrient trapping and possible enrichment of both marsh and littoral zones.

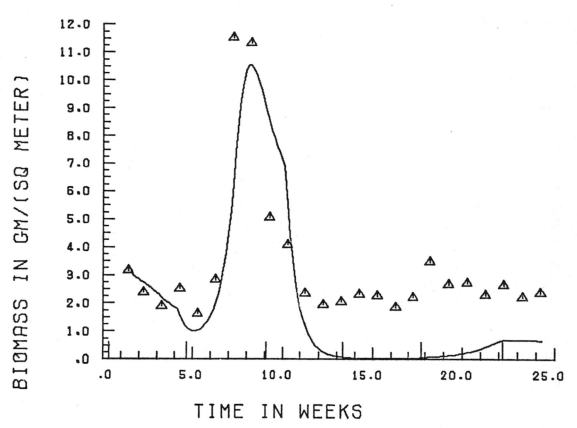

Figure 21. Open water model output for phytoplankton. Simulations for multiple species phytoplankton association with observed nutrient input.

51

At the time these simulations were obtained, nutrient regeneration by individual sources within the lake had not been coded. Accordingly, nutrient regeneration was assumed to be a temperature dependent constant. Estimates for nitrogen regeneration were taken from Dettmann (1973), and the range of phosphorus regeneration values was taken from actual lake values (Armstrong et al., 1972).

The resulting simulations (Figures 23 to 25) indicate that the postulated range of nutrient addition does affect the standing crop of phytoplankton. The limited response to such a wide range of nutrient addition, however, reveals less short term sensitivity to runoff nutrient sources than to regeneration within the pelagic system.

The importance of nutrient regeneration to lake systems has also been suggested by other studies. Dettmann (1973), for example, found that nitrogen regeneration from sediments was a major source of nitrogen for a seasonal model of the nitrogen content of Lake Wingra. In addition, studies of phosphorus dynamics in the open water of Lake Wingra indicate rapid turnover rates for dissolved phosphorus (Armstrong et al., 1972). These observed uptake rates range as much as an order of mangitude higher than phosphorus input rates from urban runoff. Sediment regeneration for phosphorus, however, could account for less than 10 percent of the regeneration rate (Armstrong, personal communication). The implication of these simulations is, therefore, that already productive systems will be slower in responding to nutrient diversion than those with lower productivity.

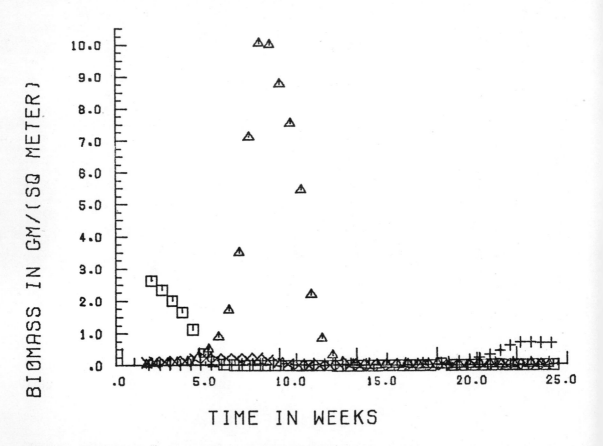

Figure 22. Phytoplankton species succession simulated in Figure 21.

Conclusions

Because of the assumptions underlying the open lake simulations, nutrient regeneration is one of the weaker model components. Future work with the model will, therefore, concentrate on more realistic representation of nutrient regeneration by the standing crop of organisms and the sediment. Nevertheless, the model in its current state does behave like the real system. It should be remembered that the model structure is founded on general principles of ecosystem functioning. Furthermore, the parameters were estimated, for the most part, independently from the simulations. Although there are insufficiencies in the model performance, the general agreement in pattern between the model and real system is an encouraging confirmation of the appropriateness of the model formulation.

In summary, this modeling effort entails the coupling of detailed mechanistic models of the hydrology of a lake's drainage basin, prediction of nutrient loading, and biotic response. Although the model is simplified for a shallow lake, system specific regressions were utilized in the simulations sparingly, and only where mechanistic models have yet to be developed. As a statement of the writers' understanding of ecosystem functioning, the model is far from perfect, but it does represent a significant synthesis of that understanding. The use of models as hypotheses of ecosystem behavior, therefore, yields a tool for studies of management alternatives for aquatic systems. As such a tool, a model helps fill the gap between scientific insight and empirical type analyses. Aquatic ecosystems are complex systems. Simulation models, while not encompassing all of that complexity, may nevertheless allow for more rigorous testing of

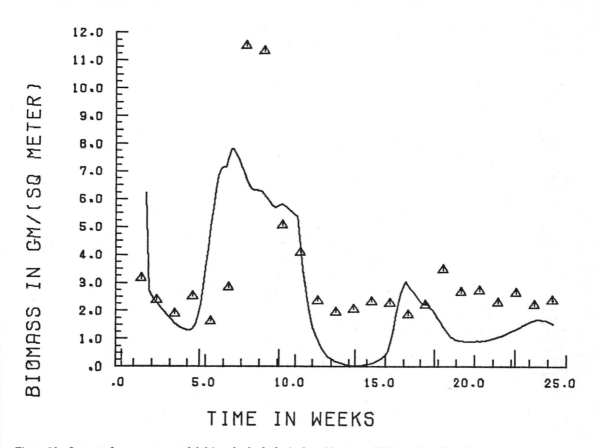

Figure 23. Output of open water model driven by hydrologic data. Nutrient addition reduced by 50 percent.

empirical models or proposed management recommendations. At the very least, this modeling effort has indicated the feasibility of using detailed simulation models of lakes and their drainage basins.

References

Allen, T. F. S. and J. F. Koonce. 1973. Multivariate approaches to algal strategems and tactics in systems analysis of phytoplankton. Ecology 54(6):1234-1246.

Armstrong, D. E., M. G. Rodel, and R. T. Bannerman. 1972. Quantification of the phosphorus cycle in Lake Wingra. Progress report. EDFB Memo Report No. 72-111. 16 p.

Crawford, N. H., and R. K. Linsley. 1966. Digital simulation in hydrology: Stanford watershed Model IV. Tech. Rept. No. 39, Stanford University.

Cullen, R. S., and D. D. Huff. 1972. Determination of land use categories in the Lake Wingra basin. EDFB Memo Report No. 72-43. 17 p.

Dettmann, E. H. 1973. A model of seasonal changes in the nitrogen content of lake water. Vol. II, Proceedings of the 1973 Summer Computer Simulation Conference, Montreal, Quebec, Canada, July 1973. pp. 753-761.

Harris, R. F., J. C. Ryden, and J. K. Syers. 1972. Phosphorus transport and mobility in land water systems. EDFB Memo Report No. 72-99.

Huddleston, J. H. 1972. Perturbation of soils in the Manitou Way residential watershed, Lake Wingra basin, Madison, Wisconsin. EDFB Memo Report No. 72-101. 37 p.

Huddleston, J. H., R. J. Luxmoore, and F. D. Hole. 1972. Soil and landscape characteristics of hydrologic response units in the Lake Wingra basin. EDFB Memo Report No. 72-100. 17 p.

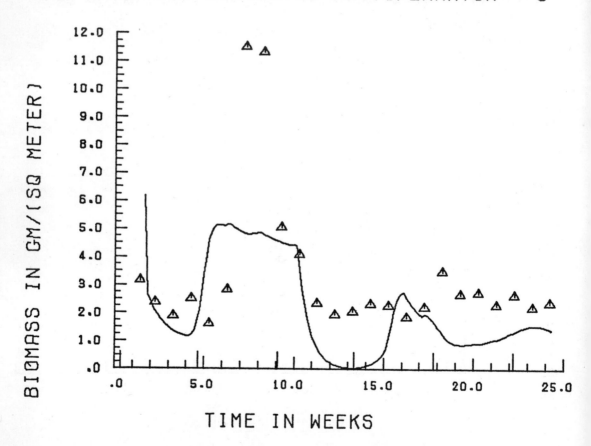

Figure 24. Output of open water model driven by hydrologic data. Nutrient input reduced by 95 percent.

Huff, D. D. 1968. Simulation of the hydrologic transport of radioactive aerosols. Ph.D. Dissertation, Stanford University.

Hutchinson, G.E. 1957. A treatise on limnology. John Wiley and Sons, Inc., New York.

Kluesener, J. W. 1972. Nutrient transport and transformations in Lake Wingra, Wisconsin. Ph.D. Dissertation, Water Chemistry Program, University of Wisconsin, Madison. 242 p.

Koonce, J. F., and A. D. Hasler. 1972. Phytoplankton succession and a dynamic model of algal growth and nutrient uptake. EDFB Memo Report No. 72-114. 112 p.

MacCormick, A. J. A., O. L. Loucks, J. F. Koonce, J. F. Kitchell, and P. R. Weiler. 1972. An ecosystem model for the pelagic zone of Lake Wingra. EDFB Memo Report No. 72-122. 102 p.

Park, R. A., R. V. O'Neill, J. A. Bloomfield, H. H. Shugart, R. S. Booth, J. F. Koonce, M. Adams, L. S. Clesceri, H. M. Colon, E.H. Dettmann, J. Hoopes, D.D. Huff, S. Katz, J. F. Kitchell, R.C. Kohberger, E. J. La Row, D.C. McNaught, J. Peterson, D. Scavia, R.G. Stross, J. Titus, P.R. Weiler, J. W. Wilkinson, and C. S. Zahorcak. (Manuscript in preparation.) A generalized model for simulating lake ecosystems.

Roberts, W. J., and J.B. Stall. 1967. Lake evaporation in Illinois. Report of Investigation 57, Illinois State Water Survey, Urbana, Illinois.

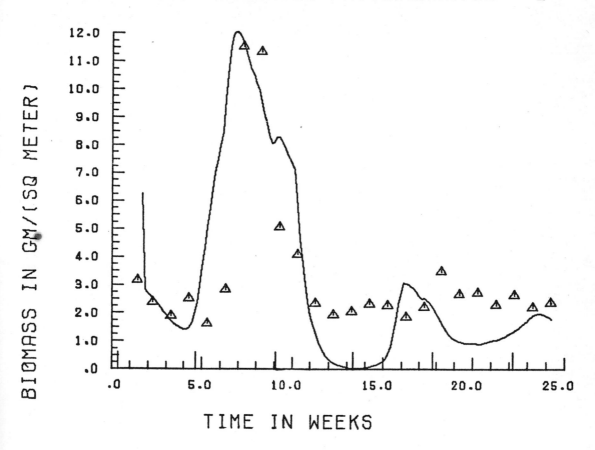

Figure 25. Output of open water model driven by hydrologic data. Nutrient input increased by 50 percent.

SOME ASPECTS OF PHOSPHORUS DYNAMICS OF
THE TWIN LAKES WATERSHED [1]

G. D. Cooke, T. N. Bhargava, M. R. McComas,
M. C. Wilson, and R. T. Heath [2]

Introduction

Eutrophication is an ecosystem-level process in which the flow of materials from the terrestrial to the aquatic portion of watershed ecosystems is greatly intensified, most often through opening of terrestrial nutrient cycles and the import of nutrients through human activities. The basis for understanding this process and for devising steps for watershed management and lake restoration is the theory of ecosystem development. Ecological succession (Odum, 1969) is an orderly and thus predictable developmental process of ecosystems which results from modification of the physical environment by the community and culminates in an ecological system in which homeostatic or stabilizing mechanisms are maximized. Mature ecosystems are ones in which development has led to, among other attributes, largely intrabiotic inorganic nutrients, closed nutrient cycles, and nutrient conservation. When a watershed is disturbed by agriculture, construction, or urbanization, mature stages are set back to developmental ones, and nutrients which were once conserved and largely intrabiotic, as well as nutrients imported by people, are added to lakes of the watershed. The result is a lake community set back in ecological time at a distance and rate related to the rate and duration of disturbance of nutrient cycles and thus loading, and which thereby exhibits characteristics of developing systems such as blooms, linear food chains, open mineral cycles, and rapid nutrient exchanges. The impact of disturbance on the lake seems to be related to lake area, volume, flushing, and to other physical and edaphic features of the watershed. These principles have been extensively elaborated by Margalef (1964, 1968) and Odum (1969), and are the basis of IBP studies on land-water interactions, such as at the Lake Wingra Basin.

The process of eutrophication, while well documented in a qualitative way, is not well understood in a quantitative sense. A description suitable for statistical purposes and for predictive models seems however to require the holistic framework of the theory of ecological succession, since the process is one which occurs at the ecosystem level of organization. A series of measurements of such processes as nutrient loading, water flushing time, and nutrient retention may thus be placed in the context of ecological time, so that together they describe developmental trends and perhaps predict the time of lake recovery. To date this has usually been possible only with systems which have a large historical data base, since the rates of these processes are normally expressed on the basis of the calendar year.

The purpose of this paper is to briefly describe a small experimental watershed which is currently undergoing nutrient (septic tank drainage) diversion, to present a sampling design which facilitates the compartmentalization of the terrestrial and aquatic portions of this ecosystem, to describe changes in water-phosphorus inflows-outflows during nutrient diversion, and to suggest a time interval for describing the eutrophication process which allows the investigator to assemble many data points for key processes over comparatively much shorter periods of time than are currently possible with the calendar year basis. It must be emphasized that these conclusions are preliminary since the writers have less than two years of work, and thus only an outline of the mathematical methods are provided here.

The Twin Lakes Watershed (TLW), located in Northeastern Ohio near Kent, is an ideal experimental ecological system. The watershed is described in detail elsewhere (Cooke and Kennedy, 1970; Cooke et al., 1973), and the essential features are summarized in Table 1. The TLW is small (209 ha) and its area is evenly divided between marsh and lakes, woodland and open fields, and urbanized area (Figure 1). There are about 300 homes on the watershed, all of which obtain water from deep wells, and until mid-1973, discharged sewage to septic tanks. There are no industries or agriculture on the watershed.

The two lowermost and largest lakes on the watershed, the Twin Lakes, are eutrophic kettle lakes of similar morphology (Figure 2 and Table 1). West Twin (WTL)

[1]This research was supported by Grant 16010HCS (R-801936), U.S. Environmental Protection Agency, and by the Center for Urban Regionalism and Departments of Biological Sciences, Geology and Mathematics at Kent State University.

[2]G.D. Cooke, T.N. Bhargava, M.R. McComas, M.C. Wilson, and R.T. Heath are with the Center for Urban Regionalism, Kent State University, Kent, Ohio.

drains slowly into East Twin (ETL), which then discharges out of the watershed. Water which enters the watershed is presumably lost either by evapotranspiration or drains into these lakes and out this outlet. Thermal stratification of the lakes occurs in late April, followed by a rapid oxygen depletion so that by mid-summer no dissolved oxygen can be found below the 6 meter depth. Autumnal circulation occurs in late October. Table 2 is a summary of the limnological features of the lakes.

Table 1. Morphological data of the Twin Lakes.

	East Twin	West Twin
Maximum Depth (M)	12	11.5
Area (ha)	26.8	34.02
Volume (M^3)	13.38×10^5	14.8×10^5
Mean Depth (M)	5.03	4.34
Maximum Length (Km)	0.85	0.65
Maximum Width (Km)	0.50	0.60
Elevation (M)	318.42	318.73

The Twin Lakes project was established in November, 1971, to study the effects of nutrient diversion on hydrologic, geochemical, and limnological features of the watershed and the Twin Lakes, and to examine methods of phosphorus inactivation as a step to reclaiming the lakes. The watershed is ideal for this purpose since it is small, there have been some pre-diversion studies (Cooke and Kennedy, 1970; Long, 1971; Heinz, 1971), and the existence of two lakes offers the possibility of a field experiment in which one lake is treated while the other is left as a control.

An essential feature of this project is the establishment of compartment models to describe the dynamics of phosphorus in the watershed and the lakes in order to (a) determine the phosphorus loading and lake retention and flushing process during and after diversion, and (b) predict what may happen if the pelagic zone is stripped of phosphorus and the profundal sediment compartment sealed by a phosphorus precipitating agent.

The first step in quantifying the eutrophication process is to study the dynamics of water and nutrient loading from the surrounding watershed to the lake in an attempt to measure where and why, and to what extent

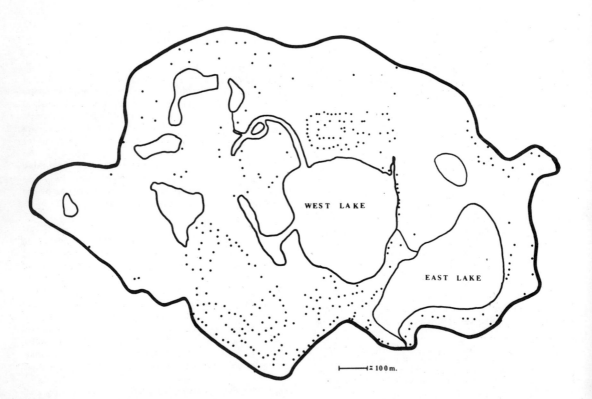

Figure 1. The Twin Lakes watershed.

58

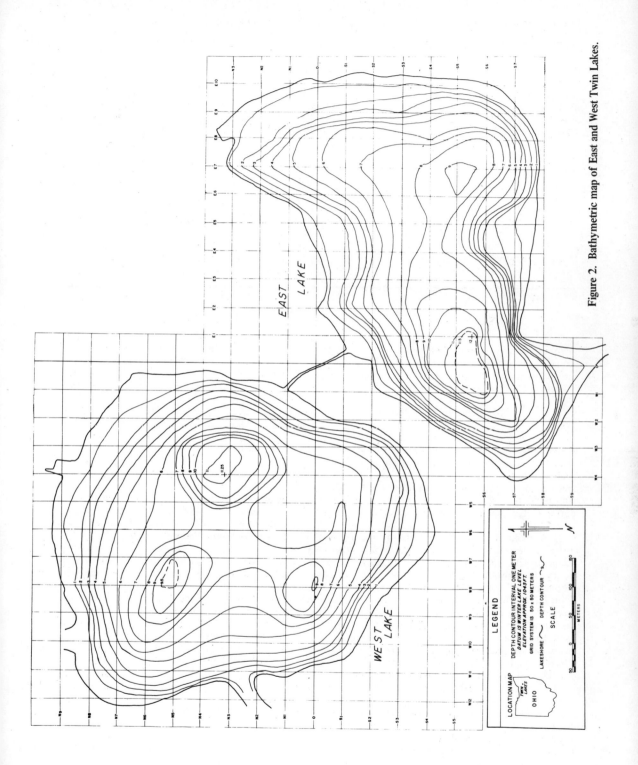

Figure 2. Bathymetric map of East and West Twin Lakes.

Table 2. Limnological features of Twin Lakes, annual (1972) minimum-maximum (0.1M depth).

	East Twin Lake	West Twin Lake
Total PO_4-P	3-325 ppb	17-330 ppb
Ortho PO_4-P	< 1-245 ppb	< 1-102 ppb
NO_3-N	< 1-187 ppb	< 1-579 ppb
CHL_A	4.8-49.1 mg/m^3	3.8-81.8 mg/m^3
Total Alkalinity (1973)	120 mg/l $CaCO_3$	120 mg/l $CaCO_3$
pH (1973)	8-9	8-9
Conductivity	2.5-4.0×10^2 μ MHOS	3.3-4.8×10^2 μ MHOS
Seston	.4-14 mg/L	0.4-13 mg/L
Plankton Volume	0.5-14.6 μl/l	0.3-31.4 μl/l
Transparency	0.8-4.6M	0.6-5.5 M
Phytoplankton (both lakes)	*Aphanizomenon flos-aquae, Anabaena limnetica, Asterionella formosa, Oscillatoria rubescens*	
Macrophytes (both lakes)	*Nuphar, Ceratophyllum, Chara, Najas, Elodea, Potamogeton*	

nutrient cycles have become opened. Management and restoration involve closing these cycles on the watershed, and perhaps in the lake as well through sealing, aeration, precipitation, or a combination of these methods.

Materials and Methods

A "grid" system for geological and limnological sampling was rejected because it presented difficulties in obtaining exact locations while in the field and because such a system would give sampling sites for the terrestrial and aquatic portions of the watershed which might not coordinate these two subsystems and thus might obscure, during analysis, their close coupling.

A spoke design (Figures 3 and 4) was developed, using the deep hole of each lake as the center, rather than the geometric center, since these points had already been in use for previous studies and could thus be called "limnological" centers. The actual circumference of East Twin Lake (ETL) was obtained from the map by placing a piece of string around the shore outline, and then dividing the string into 32 equal parts. The string was rotated on itself to obtain antipodal north and south points, then antipodal east and west points. The other diameters were selected to give a total of 8 diameters (16 spokes) that left out the least number of points on the perimeter. The radii for ETL were used in West Twin Lake (WTL) with the exception of two diameters, one used exclusively for ETL and the other used exclusively for WTL. The exception was due to the basic difference in the shapes of the lakes, but the general design allows directional uniformity in both lakes (Bhargava and Wilson, 1973).

One of the main advantages of this design is that it allows one to divide the watershed into sections and to assess the impact of changes within each section, such as rate of sewage diversion on the ground and surface waters flowing into the lake. The spokes are marked clearly at each intersection with the shore, and the deep hole marked with a large buoy. By sighting along the spokes, and with the use of a bathymetric map, field investigators may easily locate positions so that a series of samples along each spoke in various limnological zones may be taken.

Water input to the Twin Lakes is from groundwater, small surface streams, stormwater, and precipitation on the lakes. Output is by evaporation, discharge from the outlet of WTL to ETL, and from the outlet of ETL, which carries water out of the watershed.

The spoke system was employed to establish the groundwater sampling design. Shallow wells were drilled, to an average depth of 3 meters, at the intersection of each of the 32 spokes with the shore. Pumping-in tests were made to determine soil permeability, and the hydraulic gradient was measured by the water level difference between upland wells and the wells at the shore. Soil permeabilities around the lakes were similar and two years of monitoring has revealed little change in hydraulic gradient. The cross-sectional areas of sections of the watershed, equi-distant on each side of each spoke, were determined and the shallow water discharge computed for each area.

The deep flow sytem was investigated by drilling two deep wells at WTL at depths of 7 and 14.5 meters,

West Twin Lake

Spoke Number	Angles in Degrees	1/5 Radius Length (m)
1 (N)	0°	59 m.
2	30°	49 m.
3	55°	45 m.
4	75°	38 m.
5 (E)	90°	36 m
6	107°	36 m.
7	117°	37 m.
8	147°	42 m.
9 (S)	180°	70 m.
10	210°	74 m.
11	235°	78 m.
12	255°	75 m.
13 (W)	270°	74 m.
14	287°	86 m.
15	297°	92 m.
16	327°	70 m.

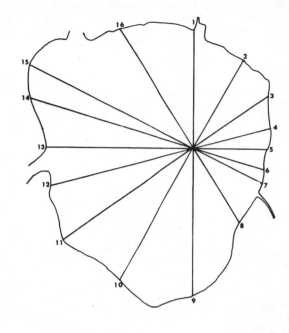

Figure 3. The spoke sampling design of West Twin Lake.

East Twin Lake

Spoke Number	Angles in Degrees	1/5 Radius Length (m)
1 (N)	0°	36 m.
2	30°	61 m.
3	42°	85 m.
4	55°	122 m.
5	75°	110 m.
6 (E)	90°	104 m.
7	107°	76 m.
8	147°	20 m.
9 (S)	180°	39 m.
10	210°	37 m.
11	222°	37 m.
12	235°	38 m.
13	255°	35 m.
14 (W)	270°	28 m.
15	287°	25 m.
16	327°	26 m.

Figure 4. The spoke sampling design of East Twin Lake.

61

placing screen on the bottom 0.6 meters of the casing, then sealing. These wells act as large diameter piezometers, and are continuously monitored with a recorder to measure head changes. The input of deep groundwater to the lakes is estimated from the upward gradient, the permeability of deep materials, and the area.

Three small streams and an overflow stream from a series of four upland, man-made lakes, flow into WTL. No permanent streams, except the outlet from WTL, flow into ETL. Small V-notch weirs were used for measuring flow of the streams and a rectangular flume was used for the overflow stream. Discharge was measured weekly.

Stormwater inputs were determined from an analysis of the change in lake levels due to a precipitation event and the amount of water flowing out of the lakes during and immediately following the storm. Since surface water inputs are not now continuously monitored, this value includes storm flows contributed by these streams, and also includes interflow or unsaturated zone groundwater flow.

Precipitation was measured by two recording rain gages, one on the watershed, the other at Lake Rockwell 1 mile east. Chemical quality of precipitation was measured from a sequential rain sampler, mounted on an unobstructed surface just next to WTL.

Evaporation was calculated from daily temperature, wind, and solar radiation, and from a recording evaporimeter.

Output from the lakes was measured at the discharge point by a stage recorder in a stilling well. A stage-discharge rating curve for each outflow was prepared, using a current meter, and discharge computed from the relation: discharge equals velocity times cross-sectional area of the channel.

Inflow and outflow streams and groundwater were monitored weekly for changes in phosphorus and other nutrients, sulfate, chloride, conductivity, and other measures of water quality which might be influenced by septic flows. These measures plus the water inflow-outflow form the basis of the phosphorus and water budgets.

Limnological sampling was based upon the spoke design as well. The pelagic zones (epilimnion, metalimnion, hypolimnion during stratified periods) were sampled weekly in both lakes at all depths for such indicators of the degree of eutrophication as plankton, seston, potential productivity, dissolved oxygen, and various forms of phosphorus. The standing crop, productivity, and phosphorus content of littoral vegetation was sampled (Rogers, 1974) along the spokes of both lakes, at shallow, intermediate, and deep zones of the littoral, and

an extensive comparative survey, over all seasons, of the amount of phosphorus in surficial sediments was made (Lardis, 1973) again using transects dictated by the spokes so that samples were obtained from sediments in contact with littoral waters and each pelagic zone. Acid and alkaline phosphatase activities were also monitored weekly. The analysis of these studies is now in progress, but it is clear that the spoke design allows the coordination of separate terrestrial-aquatic investigations so that, for example, the influence of groundwater inflow on littoral vegetation and surficial sediment phosphorus content can be examined at a series of points around each lake. The design allows an analysis of terrestrial processes, as they affect aquatic ones and thus emphasizes that the unit of study is the watershed and that the events which occur in the lake are responses to events outside of it.

Analytical techniques are those routinely employed in limnological-geological studies. In this paper only changes in water and phosphorus are described. Hydrologic methods were described earlier; total, soluble, and ortho-phosphorus (PO_4-P) were measured, using methods described in the EPA Manual (1971). Total and soluble PO_4-P were obtained by persulfate—H_2SO_4 oxidation of unfiltered and filtered samples; ortho PO_4-P was measured on filtered samples using the ascorbic acid method.

Results and Discussion

Water and phosphorus budgets

One of the first steps in measuring the eutrophication process is to monitor water and nutrient inflows and outflows of the lake. Complete nutrient budgets are a requirement for further computations and predictions, and for compartment models developed to analyze the process.

The water and phosphorus budgets for TLW were divided into meteorological seasons (Tables 3 and 4) because budgets are greatly influenced by precipitation, leaf cover of the ground, and whether soils are frozen or thawed, as well as by such factors as home construction.

The water budgets of the two lakes differ. ETL was dominated by surface inputs, mostly from the flow out of WTL to ETL, while in WTL precipitation and deep groundwater were more important. The shallow groundwater inputs of both lakes were small and a constant fraction of total input; deep groundwater was slightly more important and more variable from season to season. The largest groundwater inputs were in summer, the smallest in fall. When seasons are compared, the highest surface inflow to either lake was in spring, the lowest was in winter.

The proportions of phosphorus from each source were more similar than for water. Surface inputs were more important in ETL than in WTL, but in both this

Table 3. Sources (by percent) of water.

Season	Surface	West Twin Lake Precipitation	Groundwater Shallow	Deep	Total Input (m³)
Winter '71-72 (D,J,F)	37.7	29.9	7.6	24.9	233913
Spring '72 (M,A,M)	44.7	33.8	5.0	16.5	360236
Summer '72 (J,J,A,S)	38.1	37.1	5.8	19.0	414192
Fall '72 (O,N)	55.9	21.1	5.4	17.7	223004
Winter '72-73 (D,J,F)	43.9	25.0	7.3	23.9	243289
Spring '73 (M,A,M)	58.5	22.1	5.4	14.0	425077

Season	Surface	East Twin Lake Precipitation	Groundwater Shallow	Deep	Total Input (m³)
Winter '71-72 (D,J,F)	70.2	13.8	3.3	12.8	377488
Spring '72 (M,A,M)	76.3	14.3	1.9	7.4	662378
Summer '72 (J,J,A,S)	64.7	21.0	2.9	11.3	575849
Fall '72 (O,N)	77.8	10.5	2.4	4.3	350929
Winter '72-73 (D,J,F)	71.5	12.7	3.3	12.6	383222
Spring '73 (M,A,M)	79.7	10.9	2.1	7.3	672490

Table 4. Sources (by percent) of phosphorus.

Season	Surface	West Twin Lake Precipitation	Groundwater Shallow	Deep	Total Input (kg)
Winter '71-72 (D,J,F)	94.4	3.1	1.0	1.4	52.6
Spring '72 (M,A,M)	87.6	7.6	1.9	2.9	26.3
Summer '72 (J,J,A,S)	83.7	9.2	4.3	2.8	33.2
Fall '72 (O,N)	79.1	7.8	7.8	5.3	13.7
Winter '72-73 (D,J,F)	69.6	10.3	10.3	9.8	13.7
Spring '73 (M,A,M)	83.8	4.7	6.9	4.6	31.7

Season	Surface	East Twin Lake Precipitation	Groundwater Shallow	Deep	Total Input (kg)
Winter '71-72 (D,J,F)	93.4	3.7	1.0	1.9	33.2
Spring '72 (M,A,M)	96.3	1.9	1.0	0.8	82.4
Summer '72 (J,J,A,S)	91.3	5.2	1.8	1.7	45.1
Fall '72 (O,N)	92.8	2.7	2.1	2.4	27.6
Winter '72-73 (D,J,F)	91.4	3.2	1.8	3.5	31.3
Spring '73 (M,A,M)	93.6	2.7	1.5	2.1	56.9

source accounted for at least 70 percent and usually, 80 to 90 percent of the total income. Precipitation was an important secondary source; input from shallow and deep groundwater was usually small and constant. Phosphorus from leach fields seems to enter surface rather than groundwater flows at Twin Lakes. The experiments suggest that at least some leach fields have a perched water table, perhaps caused by bacterial slime. These fields, built in ideal materials of coarse sand and gravel, thus have greatly reduced permeability and release phosphorus to surface drainage as seeps (R. Buller, personal communication). In terms of absolute input, seasons vary greatly and as discussed later, appear to be partly influenced by the diversion of septic inflows. As pointed out by Megard (1970) for Lake Minnetonka, phosphorus and water budgets are distinct.

The phosphorus and water budgets are considered to be preliminary for a number of reasons. First, there is as yet no measure of dry fallout, and it might be a significant source since Northeastern Ohio is a major industrial area, and because there is extensive agriculture practiced on nearby watersheds. Secondly, the total phosphorus content of surface inflows (streams and stormwaters) was measured after persulfate-sulfuric acid oxidation, rather than the perchloric acid method of Mehta et al. (1954). Thus the total phosphorus input was underestimated by failing to measure phosphorus atoms associated with materials such as clays, but those phosphorus atoms probably have been measured which could become available for plant growth. Preliminary calculations of internal loading indicate that it may be a significant source of phosphorus; the extent to which it influences surface waters is not yet known. Finally, there are no continuous recording and sampling devices on inflowing streams; estimates of water and phosphorus inflows are based upon weekly samples. The variability in these inputs has been examined and it appears that the single weekly flow measurement differs significantly from mean flow, and that, with the exception of storm events, the mean concentration of phosphorus in inflowing water does not vary over a day.

Dynamics of water and phosphorus

The inflows and outflows of water and phosphorus to and from the lakes are presented in Tables 5 and 6. The diversion of sewage has apparently had little effect on the volume of water entering the lakes. The highest inflows of water were in spring months, the highest inflows of phosphorus were in winter, spring, and summer. In ETL, the months of highest phosphorus input were the same as months of highest water input since both the water and phosphorus budgets of the lake were dominated by the surface inflow from WTL. In WTL, the periods of highest phosphorus inflow were variable, found in all seasons except summer, and usually unrelated to volume of inflow (e.g. compare January, 1972, with June, 1972).

Table 5. West Twin Lake, water and phosphorus inflow-outflow.

Month	Water (M^3)		Phosphorus (kg)	
	Input	Output	Input	Output
Nov. '71	72877	103719	9.08	8.19
Dec.	107019	115071	10.23	10.78
Jan. '72	66312	76142	20.28	7.85
Feb.	60582	60582	22.06	7.16
Mar.	132806	147926	9.03	33.40
Apr.	127653	140613	14.14	12.12
May	73857	105965	4.24	5.03
June	70569	68204	4.01	1.37
July	83970	102156	4.82	3.49
Aug.	81894	73254	5.14	1.61
Sept.	177759	130240	19.24	5.83
Oct.	99068	146587	8.77	8.75
Nov.	123936	84776	8.08	7.44
Dec.	93795	123995	5.27	12.11
Jan. '73	71139	43179	3.29	3.65
Feb.	68355	61008	5.12	5.83
Mar.	143384	164541	9.54	12.72
Apr.	131184	118224	7.31	9.67
May	150509	155509	14.81	14.31

Table 6. East Twin Lake, water and phosphorus inflow-outflow.

Month	Water (M^3)		Phosphorus (kg)	
	Input	Output	Input	Output
Nov. '71	142368	151278	12.32	5.96
Dec.	189994	189994	15.62	9.08
Jan. '72	100597	106597	9.23	8.70
Feb.	86857	76897	8.36	6.63
Mar.	226978	223018	41.12	10.87
Apr.	262980	266260	24.76	22.51
May	172420	187360	16.56	9.05
June	132030	123040	13.19	7.48
July	142772	148082	9.45	14.64
Aug.	103465	102303	7.24	3.72
Sept.	197582	163624	15.21	6.54
Oct.	172269	203419	11.39	8.46
Nov.	178660	178662	16.25	14.39
Dec.	203413	194733	18.63	12.42
Jan. '73	85822	68000	5.63	4.54
Feb.	94717	84184	7.06	7.31
Mar.	250896	260605	18.59	21.39
Apr.	199218	185440	15.36	26.01
May	222376	228043	22.95	21.62

Water turnover time (lake volume divided by annual outflow) at Twin Lakes was 1.17 years (WTL) and 0.68 years (ETL); phosphorus turnover (phosphorus content divided by annual outflow) was 1.79 years (WTL) and 0.72 years (ETL), for the calendar year 1972. Both computations are based upon either mean lake volume or phosphorus content. The similarity of these figures suggests the close relationship, described in a later paragraph, between water and phosphorus outflow.

External phosphorus loadings of WTL and ETL are presented in Figures 5 and 6. Gross loading was computed by dividing total input (gms) (Tables 5 and 6) by lake area (M^2); net loading was calculated by subtracting phosphorus lost via the outflow from the total input and then dividing by area (Tables 7a and 7b). On an annual basis (1972) gross loading ($gms/m^2/yr$) of ETL was 0.712, net loading 0.466; for WTL gross loading was 0.368, net loading 0.312. These values, for lakes with mean depth of about 5 meters, are near the low side of the loading level characteristic of eutrophic lakes, as suggested by Vollenweider (1970). This indicates that following diversion the loading levels of these lakes could decline to the point of lakes classified as mesotrophic. Whether loading does exhibit a significant decline over the next years is an important question because, unlike many previously reported cases regarding the effect of diversion on the degree of eutrophication, nutrients entering the Twin Lakes do not come from point sources. This watershed may be the more typical case where nutrients originate from several sources, including now abandoned leach fields which may continue to yield nutrients for years. If loading does not drop below critical levels then will restorative techniques, such as phosphorus inactivation which may retard only internal loading from profundal sediments, be of significant value? Multiple restorative strategies, such as further reduction of external loading through treatment of stream inflow water, may be a necessity, and is consistent with the concept that lakes are parts of ecosystems with multiple sources of nutrients. Further, like many similar areas, the density of homes and the extent of additional construction may continue to leave nutrient cycles open and the watershed in an increasingly less mature state as protective zones, such as forests and marshes are lost to development.

At least in terms of lake restoration, net loading may be the more useful value with which to describe and predict successional changes following diversion. Figure 7 shows the trends in net phosphorus loading as diversion took place. Each loading peak was associated with a large water input, although not all months of heavy water inflow resulted in high net loading; amount of phosphorus lost from the lakes was, however, positively correlated with amount of water outflow ($r = .64$; $p = .01$). Thus loading of ETL was greatly dependent upon water outflow from WTL, as indicated earlier by the water and phosphorus budgets. The effect of diversion on loading and upon indicators of the degree of eutrophication of the

Table 7a. Phosphorus loading to West Twin Lake.

Month	Gross Loading ($gms/m^2/month$)	Net Loading ($gms/m^2/month$)
Nov. '71	.027	.0026
Dec.	.030	-.0016
Jan. '72	.060	.0365
Feb.	.065	.0438
Mar.	.027	-.0716
Apr.	.042	.0059
May	.012	.0023
June	.012	.0078
July	.014	.0039
Aug.	.015	.010
Sept.	.057	.039
Oct.	.026	.00008
Nov.	.024	.0019
Dec.	.015	-.020
Jan. '73	.0106	-.001
Feb.	.015	-.0003
Mar.	.028	-.009
Apr.	.021	-.007
May	.044	.0014

Table 7b. Phosphorus loading to East Twin Lake.

Month	Gross Loading ($gms/m^2/month$)	Net Loading ($gms/m^2/month$)
Nov. '71	.046	.0237
Dec.	.058	.0244
Jan. '72	.034	.0020
Feb.	.031	.0064
Mar.	.153	.1125
Apr.	.092	.0084
May	.062	.0280
June	.049	.0220
July	.035	-.0193
Aug.	.027	.0131
Sept.	.057	.0322
Oct.	.042	.0109
Nov.	.060	.0069
Dec.	.069	.0231
Jan. '73	.021	.0040
Feb.	.026	-.0009
Mar.	.069	-.0104
Apr.	.057	-.0396
May	.085	.0049

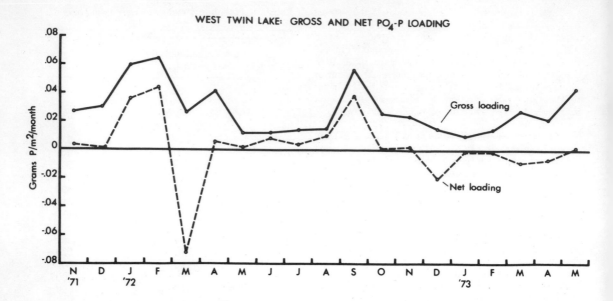

Figure 5. Gross and net PO_4-P loading of West Twin Lake.

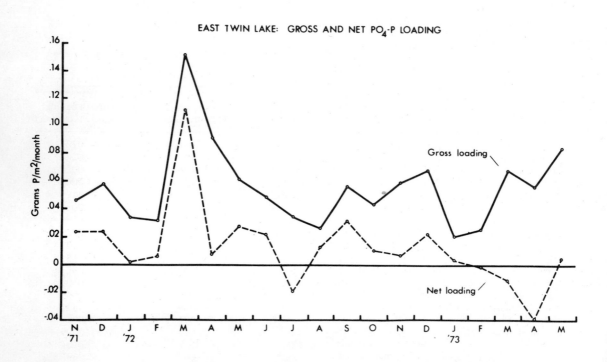

Figure 6. Gross and net PO_4-P loading of East Twin Lake.

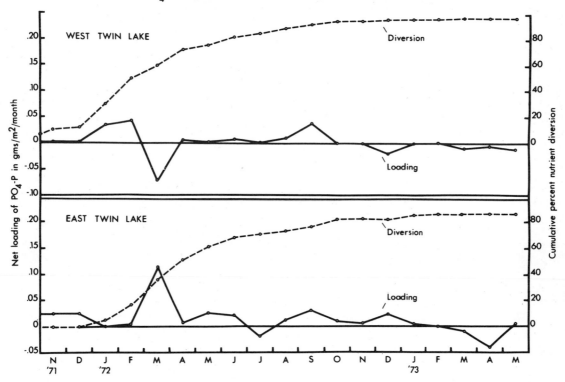

NET PO$_4$-P LOADING AND CUMULATIVE PERCENT NUTRIENT DIVERSION

Figure 7. Net PO$_4$-P loading and sewage diversion of East and West Twin Lakes.

lakes may only be better understood when more observations are available. The difficulty of this analysis will be compounded by a lack of long-term pre-diversion data. Both lakes, however, seem to exhibit a trend toward negative loading, although it is not clear whether this is due to phosphorus diversion, to increased outflows, or to other factors. When estimates of internal loading are available these values will be added to or subtracted from the net external loading to give actual loading.

On a calendar year basis (1972) inputs of neither water nor phosphorus equalled outputs. WTL lost 68239 m^3 of water but gained 18.92 kg (15.12 percent of input) of phosphorus, while in ETL, 6028 m^3 of water were gained and 65.98 kg (34.47 percent of input) of phosphorus were retained. The fraction of the phosphorus input which is retained (or lost) varied widely from month to month (Table 8) with an apparent trend toward more lost than entered during spring 1973, a period when diversion was nearly complete. The fraction of the input retained is the sedimentation coefficient (Vollenweider, 1969; Emery et al., 1973) and in ETL, varied between 0.06 and 0.74, and in WTL, varied between 0.002 and 0.71. In each lake however, there were several months when more went out than in and sedimentation was zero.

The amount of phosphorus as both ortho PO$_4$-P and total PO$_4$-P in the lakes during 1972 varied from a low in early May to a high in late October just before thermal stratification ended. Low phosphorus values were associated with mixing and high oxygen content throughout, high phosphorus with thermal stratification and presumably low redox potentials. The range of phosphorus content was sufficiently wide (e.g. in ETL in 1972, the total PO$_4$-P in lake waters ranged from about 60-300 kg) to make the computation of a mean unjustified; phosphorus turnover time, reported in an earlier paragraph, may have little meaning for the Twin Lakes when computed on an annual basis.

Eutrophication models and ecological time

A recent trend in limnological research has been the calculation of annual water and phosphorus residence times and from these and other data the development of models to predict lake recovery following nutrient diver-

Table 8. Fraction of input phosphorus retained (sedimentation coefficient; < 1.00) or lost (> 1.00).

Month	West Twin Lake	East Twin Lake
Nov. '71	.10	.52
Dec.	1.05	.42
Jan. '72	.61	.06
Feb.	.68	.21
Mar.	3.70	.74
Apr.	.14	.09
May	1.19	.45
June	.66	.43
July	.28	1.55
Aug.	.69	.49
Sept.	.71	.57
Oct.	.002	.26
Nov.	.08	.11
Dec.	2.30	.33
Jan. '73	1.11	.19
Feb.	1.14	1.04
Mar.	1.33	1.15
Apr.	1.33	1.69
May	.03	.06

sion (e.g. Dingman and Johnson, 1971; Emery et al., 1973). A number of assumptions are implicit in these models, as discussed by Sonzogni and Lee (1972). The exponential decay model based upon water residence time (years) assumes complete mixing of phosphorus throughout the lake, no net change in lake volume, constancy of phosphorus input from sources other than sewage, and also assumes that phosphorus behaves as a conservative element. Models based upon phosphorus residence time, such as that of Sonzogni and Lee (1972), may account for processes which affect phosphorus, but adopt a mean annual steady state phosphorus content and assume the mean phosphorus residence time to be a constant. Vollenweider's (1969) model, as discussed by Emery et al. (1973), also takes into account the biological and chemical processes in the lake which affect phosphorus content by employing a mean sedimentation coefficient. In all of these models the calendar year is assumed to be an actual unit of ecological time and lake recovery is equated with a reduction in phosphorus content.

Based upon data to date, these assumptions appear not to apply well to the Twin Lakes Watershed. The rates of inflows, sedimentation of phosphorus, and outflows vary widely from month to month, with consequent changes in lake water storage and phosphorus content. The use of averages would lead to weak predictions. The amount of phosphorus is further influenced by internal loading during thermal stratification. There seems now to be no basis for assuming that there is a "steady state" phosphorus content or a constant phosphorus residence

time in this rapidly evolving system. Such models as described above may not be at all applicable to small watersheds with small lakes of low mean depth because the impact of nutrients from processes of seemingly small magnitude, such as the construction of a home or a heavy storm, may be very large on water and phosphorus budgets. Larger watersheds, with more protective areas and larger lakes may respond more slowly to such processes, or at least have less frequent or rapid oscillations. The impact of successional events on the lake may be in part a function of area. These models are based on systems with point sources of nutrients which dominate the nutrient budget. The watershed in these cases may be of sufficient area to contain a diversity of habitats, including large mature areas, so that reduction of the point source alone will drastically and permanently alter the nutrient budget and thus allow a new "equilibrium" between input, storage, and output. Small watersheds with a high and increasing density of people, as at Twin Lakes, supply nutrients from a variety of sources to the lakes in addition to septic tank drainage, and may have comparatively much smaller protective or mature areas. Thus lake loading per unit time may be highly variable since it is closely coupled to watershed activities, the amount of protective area, and the rate of change (succession) in these areas.

Can lake recovery be predicted through a model which states the time when the phosphorus content of the lake will be reduced to some arbitrary value? There is a well-known relationship between some biological indicators of eutrophication, and the amount of phosphorus, at least in moderately hard-water lakes. Yet Emery et al. (1973) report a delayed recovery of Lake Sammamish, compared to the time predicted by Vollenweider's phosphorus-based model, and suggest the delay to be due to morphometric characteristics of the lake and slow flushing. Lakes with lower mean depth and thus with a short distance between productive and regenerative layers may be expected to have a high potential for plant growth, perhaps for a long period, even though external nutrient loading has been reduced. For large, deep lakes with point sources for the majority of nutrients and a large, more stable watershed, these models may be more predictive since the majority of the productive area does not lie close to sediments and because the productive zone can become rapidly impoverished of phosphorus due to loss to outflows and to deep lying sediments. Thus it seems that, at least for shallower lakes of small area and low mean depth, where various activities on the watershed, in addition to sewage input, are significant in the nutrient budget, prediction of lake recovery will be more difficult. Perhaps for future models a quantitative answer to the question "what is recovery?" is required first and that, at least for ecosystems such as at Twin Lakes, these models of lake recovery include factors such as lake area, volume, net loading, and the ratio of terrestrial to aquatic area. In this way, lake recovery is viewed as a successional process and attributes related to ecosystem structure, nutrient cycling, and nutrient conservation are considered.

The calendar or legal year, the traditional time base for calculations and expressions of lake flushing time and lake recovery, does not seem to be the appropriate time base for calculations of certain limnological processes at Twin Lakes and the calendar year has the additional disadvantage of giving the analyst only a single data point every 365 days so that many such years may be required to actually test models. There simply may be no reason to believe that the calendar year is the actual interval of time for hydrological activities such as flushing. At Twin Lakes there is, to date, less than two calendar years of data and the writers therefore are exploring the possibility of a different time interval, the limnological year, upon which to base calculations. It is suggested that the limnological year, defined as follows, may describe an actual time period on the watershed based upon hydrologic events, and at the same time yield a continuous series of data points so that mathematical and statistical analyses of the time-dependent process of eutrophication may be greatly facilitated. At the same time more traditional approaches and time intervals are being studied and compared to the limnological year.

Nutrients enter and leave lakes almost exclusively via water transport. A limnological year (LY) is defined as the length of time (days) required for the sum of the monthly water outflows of the lake to equal the initial lake volume. The LY may be a variable number of days for successive starting times (say months) since it will be influenced by changing meteorological, hydrological, and limnological conditions, such as occur over both short-term (seasons) and long-term intervals. The computation also takes into account changes in the lake, such as lake volume and flushing rate, over short intervals since a new LY is computed each month. Short-term changes are not ignored or smoothed out (recall the wide range of values of the sedimentation coefficient) but become part of the description of the process.

Let LY_I denote the limnological year which begins at time I and is computed as follows: Let V be the volume of the lake. At time I let ΔS_I be the change in storage of the lake as evidenced by the change in the level of the lake, so that the volume of the water in the lake at time I is $V \pm \Delta S_I$; note that ΔS_I can be positive or negative. Suppose the observations are made at time $i = 1, ..., k$ and the corresponding time periods expressed in calendar days are given by $t_1, t_2, ... t_k$; and let the total outflows over these time periods be given by $O_1, O_2, ..., O_k$. Then if

$$\sum_{i=1}^{k} O_i \leq V + \Delta S_I < \sum_{i=1}^{k=1} O_i,$$

then

$$\sum_{i=1}^{k} t_i \leq LY_I < \sum_{i=1}^{k+1} t_i.$$

The actual value is obtained in an obvious manner;

$$LY_I = \sum_{i=1}^{k} t_i$$

if the left hand equality for outputs is met, and it takes a properly interpolated value between

$$\sum_{i=1}^{k} t_i \quad \text{and}$$

$$\sum_{i=t}^{k+1} t_i \quad \text{otherwise.}$$

Table 9 lists the limnological years for ETL and WTL. Thus, in November, 1971, the volume of ETL was $1,345,447$ M^3. The first LY is the sum of the outflows of this and succeeding months, including fractions of months, which equal this volume. In this LY it took 248 days for the lake to discharge a volume equal to the initial volume. A new LY was computed then for each month up through present lake discharges. Each new LY accounts for changes in lake volume and outflow.

The duration of each LY seems to be fairly constant, particularly in WTL (ETL range 236-281 days; WTL range 422-442 days), even through the LY's span months which are different hydrologically, meteorologically, and limnologically (e.g. water inputs, Tables 5 and 6). Further observations will establish whether this constancy of the LY is an actual feature of the watershed.

Table 9 also lists phosphorus flushing time, in LY. This value is obtained in the same manner as the LY, by summing phosphorus outputs from each initial time, I, until the starting amount of phosphorus in the lake has flushed out. A new value is calculated for each month which takes into account changes in phosphorus content from loading and loss during the preceding month. These values are more variable than the LY, particularly in ETL, and seem to be unrelated to the duration of the LY (compare phosphorus flushing times of LY's of about 240 days) or to the initial amount of phosphorus. Further observations are required to establish the significance of these values and to establish whether there are trends related to changes in the watershed and lakes.

As indicated earlier, this concept is being examined because traditional time bases yield only a few data points over very long periods, tend to smooth out variations which may be of considerable limnological significance, and may in fact not represent an actual ecological time interval for certain limnological events related to the eutrophication process. The calculation of the limnological year yields continuous data points, after the first year has passed, which represent actual discharge rates of the lakes, a factor clearly related to changes in the lakes and watershed. Thus, mathematical and statistical analyses are facilitated and certain limnological events, measured in intervals of time related to actual ecological processes, may be more clearly portrayed. Further, the development of predictive models, such as those suggested earlier, may be greatly enhanced by the inclusion of this time base.

Table 9. Limnological years (LY) and phosphorus flushing time.

West Twin Lake

Year & Interval	Days	P Flushing (in LY's)
1 1 Nov. '71 - 27 Dec. '72	422	1.45
2 1 Dec. '71 - 5 Feb. '73	433	1.44
3 1 Jan. '72 - 18 Apr. '73	442	1.57
4 1 Feb. '72 - 31 Mar. '73	425	1.75
5 1 Mar. '72 - 10 Jun. '73	434	1.51

East Twin Lake

Year & Interval	Days	P Flushing (in LY's)
1 1 Nov. '71 - 6 Jul. '72	248	1.33
2 1 Dec. '71 - 6 Aug. '72	249	1.01
3 1 Jan. '72 - 22 Sep. '72	265	1.45
4 1 Feb. '72 - 9 Oct. '72	251	1.21
5 1 Mar. '72 - 22 Oct. '72	236	1.12
6 1 Apr. '72 - 26 Nov. '72	240	1.45
7 1 May '72 - 19 Jan. '73	264	1.15
8 1 Jun. '72 - 8 Mar. '73	281	0.95
9 1 Jul. '72 - 25 Mar. '73	268	1.30
10 1 Aug. '72 - 14 Apr. '73	257	1.19
11 1 Sep. '72 - 1 May '73	243	1.25
12 1 Oct. '72 - 28 May '73	240	0.54

Analysis of phosphorus dynamics

A compartment model based upon the premises that the eutrophication process is an ecosystem level one and that restoration and management will necessarily have to occur at this same level, has been adopted to examine the effects of nutrient diversion on the lakes and to investigate the efficacy of phosphorus inactivation as a restorative procedure following diversion. At this time insufficient data are available for presentation. In this model the totality of interactions between various components is expressed by means of a system of ordinary linear differential equations:

$$\frac{dM_i}{dt} = \sum_{k=1}^{n} a_{ik}(t) M_k(t) + F_i(t), \quad i = 1, 2, \ldots n$$

where M_i's denote mass of compartments C_i, $i = 1, 2, \ldots$, n; the time-dependent coefficient a_{ik}'s reflect the change per unit time in the value of parameter M_i due to the presence of the quantity M_k from the compartment k; F_i's denote the change per time in the value of the parameter M_i, $i = 1, 2, \ldots, n$, as caused by the application

of those external forces which are not dependent on the other parameters. Many limnological processes seem to have many aspects of nonlinearity and thus initial assumptions may be excessively simplistic and thus require considerable modification.

A compartment model of the lakes (Figure 8) based upon the work of Confer (1972), and containing the following in-lake compartments was adopted: littoral zone, pelagic zone, and sediments during unstratified periods (mid-October-April), and littoral zone, epilimnion, metalimnion, hypolimnion, aerobic sediments and anaerobic sediments during stratified periods (May - mid-October). The boundaries of the pelagic compartments are determined by temperature measurements.

The initial objective is to determine the rate and amount of contribution of phosphorus from each of the watershed compartments (surface and groundwater inflows, precipitation, and fallout) to the lake and to determine the amount of phosphorus in each of the lake's compartments and the transfer rates between these compartments. The goal is to describe the process of phosphorus loading, as affected by sewage diversion, and to describe the dynamics of this element in the lake. A study of the input-output process in relation to flow analyses, involves inputs which externally affect the system, chemical processes which internally affect the system, and outputs which account for materials lost. For a dynamic analysis of such a process flow functions are described in terms of fixed designated units of time by the system of differential equations. Transfers of specified energy or mass in terms of ratios a_{ik} are represented by means of transition matrices for the entire ecosystem. This model, which is expressed by a stochastic, possibly Markovian process, helps provide answers to various questions on transient response.

Wherever possible, transfer rates between compartments are being obtained directly through field and laboratory experimentation. However, the design and completion of such work is a lengthy process and the application of the results to mathematical analysis and then extrapolation to rates in nature is difficult. Predictions and evaluations of restorative methods in a short span of time are needed, if lake restoration through phosphorus precipitation or inactivation is to take place within any reasonable period. Thus reliance on published transfer rates may be heavy.

Ultimately the writers hope to be able to simulate the effect of various reductions in phosphorus loading, both of the external type through sewage diversion and internal type through phosphorus inactivation. Predictions made by using simulation can be tested against actual future observations and model equations can be changed.

PRELIMINARY PHOSPHORUS MODEL

(STRATIFIED LAKE)

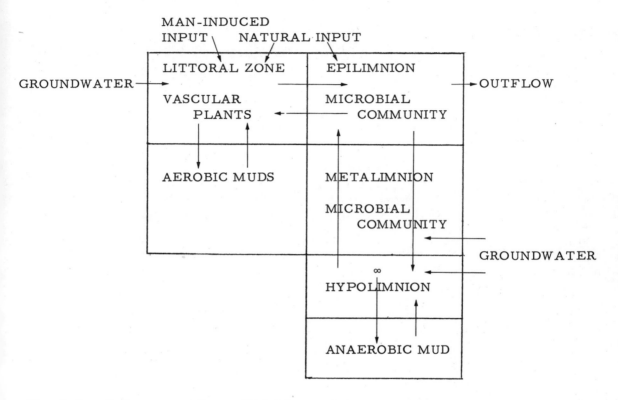

Figure 8. Generalized compartment diagram of Twin Lakes watershed.

Summary

Eutrophication is a successional process and the theory of ecosystem development provides the holistic framework needed to understand the relationship between activities on the terrestrial portion of a watershed and the response of aquatic communities to changes in these activities. In this paper an experimental watershed, the Twin Lakes Watershed, and a sampling design which greatly aids in an analysis of this ecosystem are described.

Water and phosphorus inflows and outflows of the two lakes have been measured, along with many other geological and limnological aspects not reported here, since November, 1971, just prior to sewage diversion. The water budgets differ in that East Twin input was dominated by outflow from West Twin. In both, ground-water inputs seemed to be a comparatively small and constant fraction of the total. The most important source of phosphorus was surface inflows; groundwater inputs of phosphorus were small and fairly constant.

Gross and net external phosphorus loading $(gms/m^2/yr)$ of the lakes were those characteristic of eutrophic lakes. Trends in net loading following diversion appear to be toward negative loading, but there is some indication that multiple strategies of lake restoration may be required where nutrient sources are more diffuse.

The assumptions normally applied in order to make predictions regarding the time of lake recovery seem not to be justified for the Twin Lakes. Further the calendar or legal year as an interval for calculations may not represent an interval of limnological time, thus for our data we are exploring a new time interval, the limnological year, which is based on the duration of an ecological event and which may greatly facilitate statistical and mathematical analyses.

An outline of this approach to modeling phosphorus dynamics and lake restoration is presented.

Literature Cited

Bhargava, T.N., and M.C. Wilson. 1973. On some problems in statistical analysis of an aquatic ecosystem. 39th Session of the International Statistical Institute, Vienna, August 1973.

Buller, R. 1973. M.S. Thesis, Department of Geology, Kent State University. (Personal communication.)

Confer, J.L. 1972. Interrelations among plankton, attached algae, and the phosphorus cycle in artificial open systems. Ecol. Monogr. 42:1-23.

Cooke, G.D., and R.L. Kennedy. 1970. Eutrophication of Northeastern Ohio lakes. I. Introduction morphometry, and certain physico-chemical data of Dollar Lake, Ohio. J. Sci. 70:150-161.

Cooke, G.D., M. McComas, T.N. Bhargava, and R. Heath. 1973. Monitoring and nutrient inactivation studies on two glacial lakes (Ohio) before and after nutrient diversion. Interim Research Report. Center for Urban Regionalism, Kent State University.

Dingman, S.L., and A.H. Johnson. 1971. Pollution potential of some New Hampshire lakes. Wat. Res. Res. 7:1208-1215.

Emery, R.M., C.E. Moon, and E.B. Welch. 1973. Delayed recovery of a mesotrophic lake after nutrient diversion. Jour. Wat. Poll. Control Fed. 45:913-925.

Heinz, M.H.E.F. 1971. A limnological study of the Twin Lakes, Portage County, Ohio; the annual variations of microcrustacea and physical, chemical, and biological parameters. M.S. Thesis, Kent State University.

Lardis, A.E. 1973. A comparison of the seasonal distribution of phosphorus in the sediments of two eutrophic lakes, Portage County, Ohio. M.S. Thesis, Kent State University.

Long, E.B. 1971. Biological and physical evidence of eutrophication in an Ohio lake. M.S. Thesis, Kent State University.

Margalef, R. 1964. Correspondence between the classic types of lakes and the structural and dynamic properties of their populations. Verh. Int. Ver. Limnol. 15:169-175.

Margalef, R. 1968. Perspectives in ecological theory. Univ. of Chicago Press, Chicago, Illinois, 111 p.

Megard, R.O. 1970. Lake Minnetonka: Nutrients, nutrient abatement, and the photosynthetic system of the phytoplankton. Interim Rept. No. 7. Limnological Research Center, University of Minnesota.

Mehta, N.C., J.O. Legg, C.A.I. Goring, and C.A. Black. 1954. Determination of organic phosphorus in soils. I. Extraction methods. Soil Sci. Amer. Proc. 18:443-449.

Odum, E.P. 1969. The strategy of ecosystem development. Science 164:262-270.

Rogers, W. 1974. Productivity study and phosphorus analysis of the macrophytes in two eutrophic lakes in Northeastern Ohio. M.S. Thesis, Kent State University.

Sonzogni, W.C., and G.F. Lee. 1972. Effect of diversion of domestic waste waters on phosphorus content and eutrophication of the Madison lakes. 73rd National Meeting, American Institute of Chemical Engineers, Minneapolis, Minnesota.

U.S. Environmental Protection Agency. 1971. Methods for chemical analysis of water and wastes.

Vollenweider, R.A. 1969. Possibilities and limits of elementary models concerning the budget of substances in lakes. Arch. Hydrobiol. 66:1-36. (Ger.)

Vollenweider, R.A. 1970. Scientific fundamentals of the eutrophication of lakes and flowing waters, with particular reference to nitrogen and phosphorus as factors in eutrophication. Organization for Economic Cooperation and Development. Paris.

THE ROLE OF THERMOCLINE MIGRATION IN REGULATING ALGAL BLOOMS [1]

R. E. Stauffer and G. F. Lee [2]

Introduction

Previous hydrological studies of eutrophic Lake Mendota, Wisconsin, have shown that the allochthonous phosphorus loading is small during June 1 to October 1, when massive, recurrent blooms of bluegreen algae cause serious deterioration of water quality. Consequently, it was hypothesized that the surges of algae biomass were dependent on refluxing of nutrients (N and P) by periodic depression of the thermocline by meteorologic inputs. Mortimer (1969) suggested that such a mechanism might be significant. This study reports the results of a field investigation of the possible significance of this mechanism as a factor controlling bluegreen algal blooms in Lake Mendota, Wisconsin. The approach used was to determine metalimnetic and epilimnetic ammonia, phosphorus and chlorophyll *a* concentrations, and monitor the lake's thermal structure in order to provide estimates of the vertical flux of N and P caused by thermocline depression. Particular attention was given to examining the correlation between depression and subsequent increases in chlorophyll for Lake Mendota in 1971.

Experimental Approach

Figure 1 is a schematic of Lake Mendota illustrating some of the details of its environment that are pertinent to this hypothesis. The prevailing wind directions in summer are shown by arrows. The hypsometry of the lake is known, which was requisite for this study. The maximum fetch is approximately 9 kilometers and occurs when the wind is out of the southwest. Meteorologic data were provided by the U.S. Weather Bureau Station at Truax Field, about 7 kilometers ENE of the lake's center. Sampling was accomplished by vertical profiling at stations approximately evenly spaced along the transects. The primary transect was parallel to the wind vector at

the time of sampling. The secondary transect was orthogonal to the first. Occasionally, it happened that the lake was becalmed, in which case stations were chosen which were widely spaced in the lake. Sampling was performed in the afternoon. Normally the afternoons are characterized by moderate southwest breezes which generate whitecaps except during cold front passage from the northwest.

Parameter estimation

Chlorophyll was sampled every meter down past the thermocline at each of 5 to 8 stations selected on any given day. Sampling took place on 24 days between mid-June and mid-October. Chlorophyll *a* concentrations were determined by fluorometry, using the filter combination reported in Strickland and Parsons (1968) on a Turner fluorometer. Water samples were filtered through Reeve Angel 934AH glass fiber filters and the glass fiber filters were extracted using a methanol solution. The solution consisted of 98 percent methanol and 2 percent bicarbonate buffer solution, by volume. The bicarbonate buffer was prepared by adding 16.8 grams of $NaHCO_3$ and 0.80 grams of NaOH per liter of water. The fluorescence units were related to 665 nm absorbances. Absorbances in 98 percent MeOH at 665 nm were related to absorbances of identical quantities of chlorophyll in 90 percent acetone. The concentrations of chlorophyll *a* were calculated following Strickland and Parsons (1968).

Several statistics connected with the temperature profiles and later used in the vertical transport model are explained in Figure 2. The *Boundary* between the epilimnion and metalimnion is taken to be that depth at which the second derivative of the T = f(z) function is a minimum. This statistic will be called D_e for depth of the epilimnion. Similarly, the boundary between the metalimnion and hypolimnion is taken to be that point where the second derivative of f is maximum. The *Thermocline* is then specified to be that point in depth below the surface where the temperature is midway between the two temperatures associated with the maximum and minimum in the second derivative. These random variables become statistics after observations of them are made on the lake. The thermocline depth is usually defined as the z value of maximum absolute value of the first derivative of the function f (Hutchinson, 1957). This is at the inflection point (where the second derivative equals zero) when the

[1] Financial support was provided under Environmental Protection Agency Research Grant No. Eutrophication 16010-EHR and Training Grant No. 5P2-WP-184-04. In addition, support was given this project by the Department of Civil and Environmental Engineering, University of Wisconsin and Department of Civil Engineering, Texas A&M University.

[2] R. E. Stauffer is with the Water Chemistry Program, University of Wisconsin, Madison, Wisconsin. G. F. Lee is with the Department of Civil Engineering, Texas A&M University, College Station, Texas.

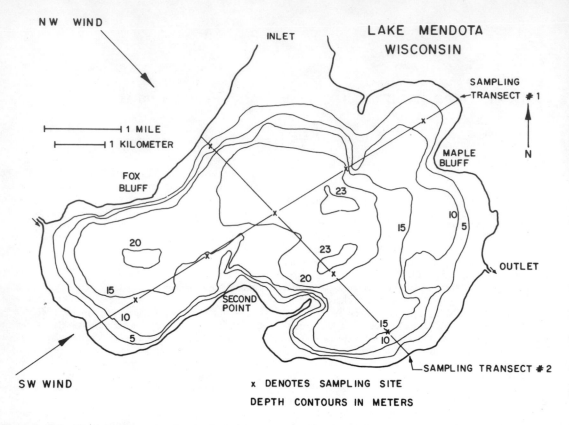

Figure 1. Lake Mendota, Wisconsin, showing bathymetry and sampling transects.

graph of f is sigmoidal. In this case, the operational definition of the thermocline explained above coincides with the classical one. Using the classical definition of the thermocline, the epilimnion boundary and thermocline are often at identical z. However, because of the steepness of the temperature curve in the upper metalimnion, a change of a couple of degrees in the temperature associated with the thermocline changes the thermocline depth by a very small amount.

Since the temperature observed at depth z at time t is a random variable which sometimes takes values in a large range over a few minutes because of progressive internal waves, and over a longer time span because of standing internal waves or active wind stress, it is not worthwhile to attach too much importance to an individual thermocline or boundary statistic. However, it is useful to graphically determine the epilimnion boundary and thermocline statistics for the various profiles on each sampling day in order to compare their values with the statistics for the average temperature curve. The average temperature curve is just the graph of the average temperature for each depth plotted against depth. This is a mean temperature profile for that day. Since the design

was to sample upwind and downwind stations with equal frequency when the lake was actively windstressed, the statistics connected with the average temperature curve are approximately unbiased.

Total phosphate followed the persulfate procedure of Menzel and Corwin (1965), using the molybdate-blue, ascorbic acid procedure in the colorimetric step. Phosphorus fractions were defined in accordance with Strickland and Parsons (1968). Ammonia-N was analyzed using an alkaline-phenol-hypochlorite AutoAnalyzer procedure as modified by Kluesener (1972). Nitrate-N analysis followed Kahn and Brezenski (1967) and also involved use of a Technicon AutoAnalyzer.

Vertical transport model

Figure 3 shows several of the mean temperature curves for the period June 16 to July 21. Tremendous heat inputs into the metalimnion and hypolimnion occurred during the period July 8-14. Note that the heat gains between 6 and 8 July and 14 and 21 July are negligible by comparison. The displacement of the

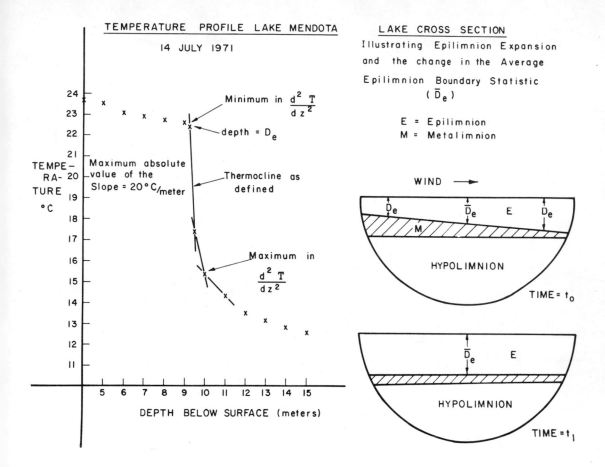

Figure 2. **Illustration of boundary statistics.**

thermocline associated with this heat transport occurring during a major cold front is typical. No significant amount of this heat gain can be accounted for by the process of oscillation of standing internal waves (Dutton and Bryson, 1962). This is so since the time interval for the heat gain is too short. On the contrary, the explanation lies in the turbulent exchange of epilimnetic water with metalimnetic water during the cold front passage. This process extends the boundary of the epilimnion with the result that nutrients in this added volume element are directly annexed by the epilimnion. In addition, it causes some warming of the water remaining in the metalimnion and hypolimnion.

If it is assumed that solutes have turbulent exchange coefficients the same as heat, then knowledge of the concentrations of ammonia and phosphorus in the metalimnion and hypolimnion prior to the wind disturbance enables the calculation of the vertical transport of nutrients using changes in average temperature profile. Approximate mixing ratios R_z between epilimnion water and the metalimnion water in each volume element at depth z in the metalimnion and hypolimnion are calculated by Equation 1. The total mass transport of solute s is the sum of the component transports from each depth element (Equation 2). Each of the component transports during the period t_1-t_0 is equal to the average excess concentration of the solute (as compared to the epilimnion) in the volume element at depth z multiplied by the volume of the element multiplied by its mixing ratio (Equation 3). The mixing ratio defined is based on conservation of heat and the constancy of the specific heat of water over the temperature range of interest. It is assumed that the epilimnion reaches a uniform temperature before epilimnion boundary migration begins. Based on the comparative coefficients of turbulence in the epilimnion and the metalimnion, and on actual field observations this is not unreasonable.

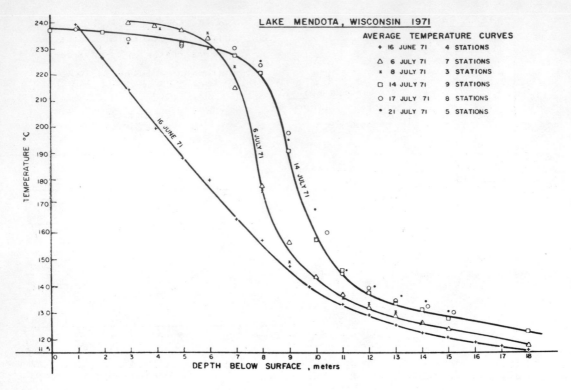

Figure 3. Lake Mendota, 1971, showing stochastic seasonal progression of average temperature curve.

Vertical Transport Model

Let:

V_j = volume of the jth depth element

$[s]_j$ = average concentration of solute s in depth interval j

$[s]_{e \cdot t_o}$ = average concentration of solute s in the epilimnion at time t_o

F_j = solute transport component from the jth depth interval

F_e = solute transport component due to epilimnion expansion

t_o = initial time

t_1 = later time

R_z = mixing ratio at depth z. Here z is the average depth of the depth interval j

$\overline{T}_{e,t_o}$ = average epilimnion temperature at time t_o

T_{z,t_1} = temperature of water at depth z, at time t_1

F = total transport of solute s over time interval (t_o -t_1)

Mass transport equations

$$R_z = \frac{T_{z,t_1} - T_{z,t_o}}{\overline{T}_{e,t_o} - T_{z,t_o}} \quad \cdots \cdots \cdots \cdots (1)$$

$$F = F_e + \sum_j F_j \quad \cdots \cdots \cdots \cdots (2)$$

$$F_j = V_j \left\{ [\overline{s}]_j - [\overline{s}]_e \right\} R_z \quad \cdots \cdots \cdots (3)$$

Let R_z be the approximate mixing ratio at depth z below the surface. R_z is approximate since it is derived considering water exchange between two water reservoirs at different temperatures, with the heat source reservoir (the epilimnion) considered to be very large relative to the metalimnion water reservoir. This assumption implies that the epilimnion temperature remains stationary in spite of the heat exchange that has taken place. In fact, the epilimnion temperature drops during a period of thermocline depression because of accelerated sensible and latent heat transfer to the atmosphere and because of water exchange with the metalimnion and expansion of the epilimnion. The cooling of the epilimnion that results from sensible and latent heat transfer to the atmosphere introduces a negative bias to R_z. Calculations indicate that this is the overwhelmingly dominant factor causing the temperature drop in the epilimnion during a cold front passage. One trouble with the mixing ratio R_z is that the average epilimnion temperature at the critical time of mixing may be imperfectly known unless temperature monitoring is possible using remote sensing. The result is that when the denominator of the R_z expression is small, the uncertainty in the ratio is large. If the epilimnion is cooled during the time interval (which is the case when deepening occurs because of cold front passages), then it is possible to define another mixing ratio statistic based on the final average temperature of the epilimnion. It is not hard to see that this new statistic R_z^* is greater than R_z for all z. In fact, R_z^* is an upper bound for the true mixing ratio. A recommended course is to use the average of R_z and R_z^* in computing transport. The uncertainty in the average epilimnion temperature at the critical time of mixing can be reduced by temperature profiling at more frequent intervals. Naturally, if the epilimnion temperature has been reduced by extensive mixing and by sensible and evaporative heat loss, it tends to rebound rapidly with the onset of still, hot weather. This happened after August 5, 1971.

The model as described predicts the transport of N and P based on measurements of the initial and final states of the density partitioned water column. There is no attempt to dynamically describe the water movements which are attendant in the meteorologic "event" which causes the mass transport. In this sense, the approach is a statistical estimation one, with the focus on determining the pertinent system parameters (such as subsystem boundaries, concentrations of solutes and heat contents), and the changes in these parameters over the time interval, as accurately and precisely as possible. No attempt has been made to separate the vertical transport into specific components such as transport resulting from upwelling or mass transport involving entrainment of colder water at the epilimnion boundary because of changes in the Richardson number accompanying the currents set up by Langmuir circulation helixes. As interesting as these individual hydraulic problems are, their exact resolution is not prerequisite to describing the net changes in the partitioning of the lake's nutrient pool over the entire time frame.

Sweers (1970) considered the thermocline as a dynamic boundary. The vertical eddy diffusivities at the bottom of the epilimnion vary over time in response to the variable inputs of momentum from windstress. The partition state of the lake is not deterministic with respect to time since the partitioning agent is the time sequence of meteorologic inputs to the lake. An effect of this is that the boundaries of the subsystems (epilimnion, metalimnion, hypolimnion) are not constant with time, even over an interval of a few days or weeks in the summer, for the subsystem boundaries respond to the continuing sequence of meteorologic inputs, the early part of which was critical in their development. Both the timing and the magnitude of the vertical nutrient transport events are random variables.

Figure 4 illustrates the chlorophyll regime before and after one of the major wind events of the summer, namely, a cold front passage accompanied by strong, enduring winds that occurred on July 12-13. The statistics plotted are the average concentrations of chlorophyll at each depth below the surface on the two successive days. One unit on the ordinate is equivalent to 0.288 mg/m^3 chlorophyll a. The bars denote the total range of the 5 order statistics for each depth on each of the two days. Using the averages and the hypsometric table, the total chlorophyll a in the lake on July 8 was 2000 kilograms and on July 14, 4300 kilograms. Nonparametric statistics can be used to test the differences between the *medians* at each depth on the two successive days. The sampling design tends to bring out the possible range of each chlorophyll statistic because upwind and downwind stations are always included. It is clear that the true power of the statistical tests comparing the median concentrations at each depth on the successive days is actually greater than can be calculated using the data presented. Nevertheless, the sample ranges at most depths do not come close to overlapping. This implies that the change in total epilimnion chlorophyll was highly significant during the period 8-14 July. Furthermore, Figure 4 indicates that only an insignificant amount of the total lake chlorophyll lies below the thermocline in Lake Mendota. This percentage decreased as the summer progressed.

As was expected from hydrodynamic considerations, the thermocline and epilimnion boundary statistics associated with the individual temperature profiles vary with distance downwind. Most importantly, it was found that for each of the two random variables (thermocline depth and epilimnion boundary), the mean of the individual statistics, the median, the average of the maximum and the minimum of the statistics, and the statistics derived from the average temperature curve usually agreed within ± 0.2 meters. The range for all stations was sometimes as great as 2 meters. The statistical estimation problem of interest is to detect *significant* displacements of the epilimnion boundary and the thermocline over time. This will be treated in detail in another paper.

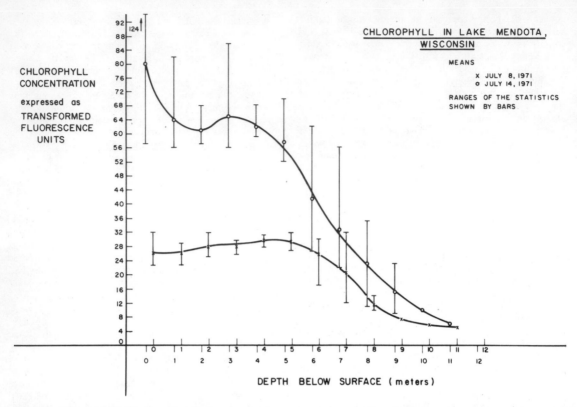

Figure 4. **Chlorophyll in Lake Mendota, 8, 14 July 1971. One unit on the ordinate is equivalent to 0.288 mg/m³** chlorophyll *a*.

It was found that the epilimnion boundary statistic is the most useful because it is usually sharply defined. It indicates expansion of the epilimnion at the expense of the metalimnion and hypolimnion, and is necessary for computing the total phosphorus in the epilimnion at any time t.

Results and Discussion

Figure 5 summarizes the observations on Lake Mendota during the stratified period, 1971. Total chlorophyll per square meter in the lake's pelagic zone (water depth > De), cumulative hours of sunshine, cumulative rainfall and the depth of the thermocline are shown.

The thermocline migrated downward 1.2 meters on July 12-13 and a total of 3.7 meters between July 12 and August 5. The thermocline migrated downward approximately a meter in the following month, i.e., between August 5 and September 9. Cumulative hours of sunshine does not present a very convincing parameter to account for the dramatic changes in chlorophyll during the summer investigated. The rainfall during the summer was seriously below normal except for the period from August 10-14, when approximately 2.5 inches of rain fell in three

thunderstorms. Consequently nutrient inputs from rain and rural runoff were probably lower than normal. The magnitude of the (chlorophyll per square meter) values attained suggests that the algae were driven into light limitation (Hepher, 1962, Steemann-Nielsen, 1962). At all times during the summer after mid-June, bluegreens were overwhelmingly dominant with *Microcystis, Aphanizomenon, Gleotrichia,* and *Anabaena* species in various combinations making up the population. *Anabaena* formed the first bluegreen algal bloom in early-mid July.

Epilimnion total phosphorus and total lake chlorophyll mass are shown in Figure 6, with the calculated vertical transport and measured changes in the total epilimnion phosphorus (mass) indicated by bars. Total chlorophyll reached very high values following those periods of active vertical transport. There are three significant maxima in the total chlorophyll curve. The first is in June when the algae were responding to inorganic nutrients in the epilimnion. This pulse came with the rapid warming of the water and development of the thermocline. Secchi disk readings throughout May and June indicate that the curve for chlorophyll was rising from late May until chlorophyll observations began on

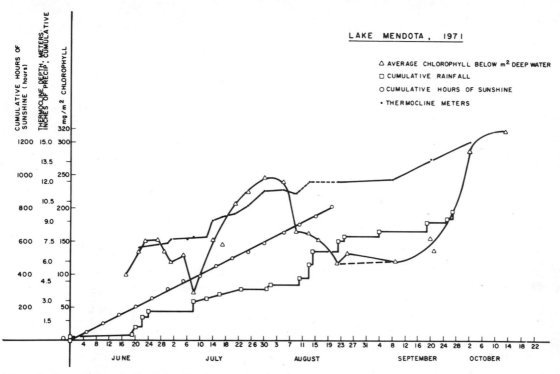

△ AVERAGE CHLOROPHYLL BELOW m² DEEP WATER
□ CUMULATIVE RAINFALL
○ CUMULATIVE HOURS OF SUNSHINE
• THERMOCLINE METERS

Figure 5. Summary graph for Lake Mendota, 1971, showing chlorophyll, cumulative rainfall, cumulative hours of sunshine, and thermocline depth as a function of time.

June 17. Exhaustion of nutrients occurred by late June in the epilimnion.

The two large chlorophyll peaks coincide with the two periods when significant thermocline migration occurred. The July period was characterized by frequent cold fronts and the September-October peak was associated with the rapid deepening that occurred with the onset of fall weather after what was initially an unseasonably warm and dry September. There was no significant amount of chlorophyll below the thermocline after the onset of sharp thermal stratification in mid-June. The minimum total chlorophyll in the lake occurred at the same time as the minimum total epilimnion phosphorus. Conversely, the corresponding maxima also agree in time. The total phosphorus content of the epilimnion increased from 5900 kilograms to 10,000 kilograms during the cold front of July 12-13. This is an increase of 70 percent. The chlorophyll more than doubled. The measured epilimnion phosphorus pool increased by 12,400 kilograms or 78 percent, between September 20 and October 2. The calculated vertical mass transport of P over the same period was 14,100 kilograms. Chlorophyll again made impressive gains. The sharp drop in chlorophyll between August 5 and August 9 was associated with an equally impressive drop in total epilimnion phosphorus and occurred during a period of hot, calm weather on the

lake. These drops can probably be attributed to settling of algae out of the epilimnion, although such a process is difficult to quantify directly.

The rapid loss of phosphorus and algae from the epilimnion during this hot, calm period in August suggests that sedimentation of a bloom can occur quite rapidly. The small increases in total lake chlorophyll around July 2 and August 25 may have their explanation in moderately strong winds that occurred on June 30-July 1 and again on August 22. The wind of June 30 thickened the epilimnion slightly but in a way that is not comparable to the big wind event of July 13. It is risky to attach too much significance to *small* changes in the total chlorophyll because chance alone may explain this variation in the statistic. This is being investigated. It must be emphasized that turbulence alone may favor development of higher total lake chlorophyll based on what is known of the physiology of photosynthesis. However, this factor does not appear to be controlling.

Nutrient transport accompanying thermocline depression can be expected to increase total lake chlorophyll only when the algae crop is limited by the nutrients being provided by this mechanism. For this reason, one would not expect vertical nutrient transport resulting from cold front passages to increase chlorophyll levels in mid-June

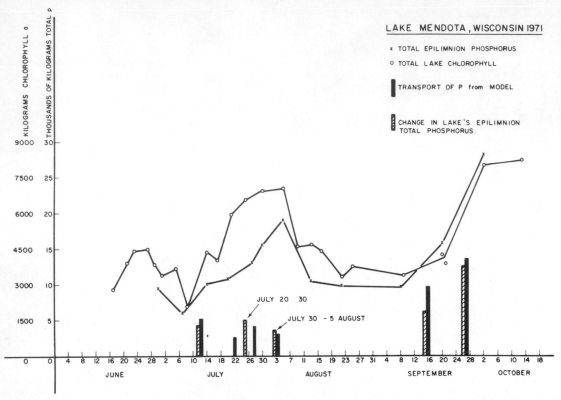

Figure 6. Lake Mendota, 1971, showing relationship between total epilimnion phosphorus and total lake chlorophyll (equivalent to total epilimnion chlorophyll for Mendota during the summer).

and mid-October when nutrient supplies in the epilimnion of Lake Mendota are ample.

It is difficult to establish cause and effect between vertical phosphorus and nitrogen transport and subsequent increases in chlorophyll. However, the following argument suggests that cause and effect can exist when the epilimnion is initially nutrient impoverished. First consider the total P content of the epilimnion on 8 July (estimated as 5900 kilos). The total is of course strictly greater than the mass of P associated with living algae biomass. In fact, data collected as part of this investigation indicates that probably less than 4000 kilos could have been part of a biomass at this time. The remaining P is in the other phosphorus fractions defined by Strickland and Parsons (1968). Furthermore, it has been estimated that live algae (marine) is approximately 0.9 percent phosphorus by weight (dry weight basis) (Redfield et al., 1963), and that algae other than the *Chlorophyta* are 0.2-1.0 percent chlorophyll *a* by weight (Strickland, 1965). Figure 6 shows that the total lake chlorophyll for Mendota reaches about 9000 kilos at times when nutrients cannot possibly be limiting. Based on the prevailing stoichiometry of algae, it seems unlikely that a 9,000-10,000 kilogram chlorophyll bloom will result

based on available phosphorus resources of about 4,000 kilograms. It must be emphasized, however, that if vertical mass transport of nutrients is *necessary* for the maintenance of large standing crops (as measured by chlorophyll) during the summer stratification period, it need not be *sufficient*.

Considering the large number of random variables involved and the blow-up in relative error that accompanies taking the difference of random variables, the agreement between changes in total P in the epilimnion as measured and the computed input because of turbulent exchange is surprisingly good. The computed transport exceeds the measured change in epilimnion P. However, the positive vertical transport component should always be greater over a time period than the net change since phosphorus loss due to sinking of algae is not considered. Sample calculations show that the P input due to the rainfalls plotted on Figure 6 is trivial in comparison to the vertical transport of P. Moreover, Rohlich (1966) reported in his review of the nutrient sources for Lake Mendota that the total P transport from streams tributary to Lake Mendota in the five-month period between May 1 and October 1, 1949, was 5900 kilograms. Forty-three percent of this figure was reactive phosphate. The total is approximately equal to the total P transport estimated to have resulted from the single cold front passage of July 12-13, 1971. In the fall, such massive amounts of P are moved up that a large reserve of reactive P accumulates in the mixed layer. This does not occur during the summer. When considering algal nutritional requirements the metalimnion of Lake Mendota is relatively rich in P as compared with N. Nitrate is unimportant in the reduced metalimnion in late summer. Ammonia reaches concentrations above 2 mg/l-N by late summer in the lower metalimnion. The mass ratio of ammonia-N/total-P calculated to have been transported vertically into the epilimnion by the mechanism outlined during the periods

indicated by the bars is approximately 3.5/l. The metalimnion and hypolimnion P is almost entirely reactive phosphorus.

Table 1 is a summary of the phosphorus and ammonia-N transport calculated using the model during 1971. Some of the figures deserve special emphasis. During the period July 12-August 5 inclusive, the vertical transport of total-P was 15,170 kilograms. For the same period the ammonia-N transport was 53,800 kilograms. For the period July 12-September 20, the figures are 25,720 and 95,390 kilos, respectively. The time period is the last two months of the summer. The total-P content

Table 1. Vertical transport of phosphorus and ammonia-N in Lake Mendota, 1971.

Date Period and Element	F-Transport	F* Transport	Average
8-14 July			
1) Total-P	5,135	5,246	5,190
2) Ammonia-N	12,753	13,029	12,891
Ratio: $\frac{\text{Ammonia-N}}{\text{Total-P}} = 2.48$			
21-25 July			
1) Total-P	2,463	2,463	2,463
2) Ammonia-N	7,581	7,581	7,581
Ratio: N/P = 3.08			
25-30 July			
1) Total-P	4,670	5,670	5,170
2) Ammonia-N	20,800	25,400	23,100
Ratio: N/P = 4.46			
30 July-5 August			
1) Total-P	2,347	2,347	2,347
2) Ammonia-N	10,260	10,260	10,260
Ratio: N/P = 4.37			
8-20 September			
1) Total-P	9,100	12,000	10,550
2) Ammonia-N	35,870	47,320	41,600
Ratio: N/P = 3.94			
20 September-2 October			
1) Total-P	13,600	14,600	14,100
2) Ammonia-N	51,570	55,360	53,470
Ratio: N/P = 3.79			
2-13 October			
1) Total-P	25,950	28,070	27,010
2) Ammonia-N	98,990	108,100	103,550
Ratio: N/P = 3.83			

Note: Transport figures are in kilograms as P and N. F and F* figures refer to the computed transport using the R and R* ratios, respectively. Linear interpolation used in estimating nutrient concentrations below the boundary De of the epilimnion as a function of time.

of the epilimnion in late August was approximately 10,000 kilos. Lake Mendota's surface area is 39.1 km^2. Hence, the transport of P/m^2 surface area between 12 July and 20 September was 658 mg P/m^2. In analyzing nutrient transport as it affects lake eutrophication, it is the "loading" or input/m^2 which is most useful in making lake comparisons.

It is important to emphasize two distinctions between phosphorus transport from external sources and phosphorus transport resulting from the turbulent entrainment of metalimnion water. First, mass transport of phosphorus calculated using the summation involving R$_z$ is entirely reactive phosphorus, since the concentration of reactive-P in the epilimnion is normally negligible. Hence, this P input is more readily available to algae than the particulate P resulting from land runoff, although this is not necessarily true for sewage-P. Secondly, the vertical mechanism described transports reactive-P directly into the non-littoral euphotic layer. Phosphorus inputs from external sources may be subjected to uptake by macrophytes or vegetation at the lake margins and therefore not become available to the phytoplankton.

It is difficult to detect small changes of temperature over time in the metalimnion and hold statistical significance. Because the means of temperatures measured at any depth j are random variables, this makes the transport model insensitive to slight perturbations of the metalimnion temperature regime. In addition, over longer time frames, downward heat flux via the mechanism of oscillating standing internal waves may play a role as suggested by Dutton and Bryson (1962). Heavy precipitation with the resulting runoff and turbidity currents can completely vitiate the model over the appropriate time frame. Bryson and Suomi (1951) discuss a dramatic case of this. Furthermore, the model is inappropriate for lakes where light penetration through the epilimnion to the metalimnion is sufficient to cause substantial warming over time (Bachmann and Goldman, 1965), This is not the situation in Mendota because of the low Secchi transparencies and the moderately thick epilimnion in summer. In spite of these sometimes important errors in the model the calculated transports over periods where these errors are negligible is impressive. Nevertheless, the mechanism outlined appears to be one control over the biomass (as measured by chlorophyll) because of its regulation of the epilimnion nutrient regime.

Conclusion

In conclusion, Lake Mendota provides an interesting example of a lake which is definitely not in a steady state with respect to nutrient transport from external sources during the stratified period. Instead, it depends on a stochastic process for the recycling of its own nutrient pool in order to maintain high chlorophyll levels during the summer. This type of periodic nutrient enrichment is likely to be important in temperature lakes where metalimnetic reserves of N and P accumulate during the stratified period and where the thermocline deepens because of storms and seasonal progression. This mechanism for recycling of nutrients must be considered in estimating intra-lake nutrient transfers and in analysis of the interactions among morphometric characteristics of lakes and the time sequence of nutrient additions from external sources as they relate to biological productivity.

References

Bachmann, R.W., and C.R. Goldman. 1965. Hypolimnetic heating in Castle Lake, California. Limnol. Oceanogr. 10:233-239.

Bryson, R. A., and V. E. Suomi. 1951. Midsummer renewal of oxygen within the hypolimnion. J. Mar. Res. 10:263-269.

Dutton, J. A., and R. A. Bryson. 1962. Heat flux in Lake Mendota. Limnol. Oceanogr. 7:80-97.

Hepher, B. 1962. Primary production in fish ponds and its application to fertilization experiments. Limnol. Oceanogr. 7:131-136.

Hutchinson, G. E. 1957. A treatise on limnology, v. 1. John Wiley, 1015 p.

Kahn, L., and F. T. Brezenski. 1967. Determination of nitrate in estuarine waters, automatic determination using a brucine method. Environmental Sci. Technol. 1:492-494.

Kluesener, J. 1972. Nutrient transport and transformations in Lake Wingra, Wisconsin. Ph.D. Thesis. Water Chemistry, Univ. of Wisc., Madison. 242 p.

Menzel, D. W., and N. Corwin. 1965. The measurement of total phosphorus in seawater based on the liberation of organically bound fractions by persulfate oxidation. Limnol. Oceanogr. 10:280-282.

Mortimer, C. H. 1969. Physical factors with bearing on eutrophication in lakes in general and in large lakes in particular. In G. A. Rohlich (ed.), Eutrophication: Causes, Consequences, Correctives. p. 340-368. National Academy of Sciences.

Redfield, A.C., B.H. Ketchum, and F.A. Richards. 1963. The influence of organisms on the composition of sea-water, p. 26-77. In M.M. Hill (ed.) The Sea V. 2, Interscience, New York.

Rohlich, G. A. 1966. Origin and quantities of plant nutrients in Lake Mendota. In David G. Frey (ed.), Limnology in North America. University of Wisconsin Press. p. 75-80.

Steemann-Nielsen, E. 1962. On the maximum quantity of plankton chlorophyll per surface unit of a lake or the sea. Int. Rev. Ges. Hydrobiol. 47:333-338.

Strickland, J.D.H. 1965. Production of organic matter in the primary stages of the marine food chain, p. 477-610. In J.P. Riley and G. Skirrow (eds.) Chemical Oceanogr. V. 1. Academic Press, London.

Strickland, J. D. H., and T. R. Parsons. 1968. A practical handbook of seawater analysis, Bulletin 167. Fis. Res. Bd. Can., 311 p.

Sweers, H. H. 1970. Vertical diffusivity coefficient in a thermocline. Limnol. Oceanogr. 15:273-281.

NEED FOR AN ECOSYSTEM PERSPECTIVE IN EUTROPHICATION MODELING[1]

B. C. Patten[2]

Introduction

In any systems modeling it is legitimate to define a system of interest and proceed to model it by determining state variables, state spaces, transition functions, etc. The defined system also implies existence of a complementary system which is everything else: the system's environment. The environment drives the system by a set of exogenous variables (inputs, forcings, loadings) which are time dependent but state independent. The system is also driven internally through the time dynamics of a set of time dependent and state independent endogenous variables (process coefficients, physical constants, etc.). Thus, a systems model consists of the interrelations between a set of state variables $x(t)$ as influenced by a set of exogenous and endogenous variables, the parameters $p(t)$.

In this symposium, the eutrophication process seems mainly to concern the primary production subsystem of lake ecosystems. The exogenous drives are basically nutrient loadings and photosynthetic energy inputs, and the internal processes include physiological parameters of algal nutrition, limnological parameters of water mass and entrained nutrient dynamics, etc. For the most part, the rest of the lacustrine ecosystem is ignored.

The purpose of this paper is to argue that the appropriate minimal conceptual unit for modeling eutrophication is the whole ecosystem. The corresponding physical unit may be a lake or watershed, as appropriate to encompass the relevant "ecosystem." The basic elements of the argument are two:

(1) Ecosystem components are all mutually dependent, conferring the system property of complete controllability. Consequently, control points for attacking eutrophication exist everywhere in the ecosystem, not just in the nutrient and producer parameters.

(2) Remote control may be more effective than direct control due to a reciprocal dependency relationship between ecosystem components. Small, correctly timed, and relatively inexpensive control actions exerted at sensitive points in the ecosystem interconnection network may be amplified in signal propagation to produce significant reversals of eutrophication states.

Mutual Dependence

Hutchinson (1948), writing on biogeochemical cycles, referred to ecosystems as "circular causal systems." Every ecosystem, regardless of how its components are defined or aggregated, has the basic dependency structure illustrated in Figure 1. Matter recycling provides the means by which the dependency loop is closed and every ecosystem, whatever its internal configuration of feedforward and feedback interactions, is in effect nested within recycle loops. The resultant closed dependency structure confers the property of mutual causality. That is, each and every state variable of an ecosystem affects every other, ultimately, as influence is propagated around the grand loop. Thus, parameter dynamics anywhere are transmitted everywhere.

A system is said to be completely state controllable if for any time t_0 it is possible to find a parameter vector $p(t)$, $t \in [t_0, t_f]$, which will transfer any given initial state $x(t_0)$ into any specified final state $x(t_f)$ in the finite time interval $[t_0, t_f]$ (Zadeh and Desoer, 1963). Loosely, controllability implies the ability of parameters to influence each state variable of the system. In ecosystems, with mutual dependence of state variables, complete controllability is the rule.

Mutual dependence of states makes the ecosystem a natural functional unit. The corresponding property of controllability makes it the conceptual unit of choice in attacking environmental degradation problems such as eutrophication. Many more potential control points are available, obviously, in the complete set of ecosystem parameters than are present in the nutrient-producer subsystem where current attention is focused. Whole ecosystem models will be required to explore the great variety of control combinations offered in such an expanded parameter set. Given the capability to construct such models, an outline for exploring alternative control combinations is given as follows.

---

[1]University of Georgia, *Contributions in Systems Ecology*, No. 14.

[2]B.C. Patten is with the Department of Zoology and Institute of Ecology, University of Georgia, Athens, Georgia.

Nominal and Perturbation Dynamics

A control action may be viewed as a perturbation of one or more system parameters. Without perturbation, the nominal dynamics of an ecosystem may be modeled by

$$\dot{x}(t) = f[x(t), p(t)], \quad x(t_0) = x_0, \quad t \in [t_0, t_1], \quad \cdot \quad \cdot \quad (1)$$

where $x(t) \in R^n$ is a vector of n mutually dependent state variables, $p(t) \in R^m$ is a vector of m mutually independent exogenous and endogenous parameters, x_0 is the initial state, and $[t_0, t_1]$ is the simulation interval. Solution of Equation 1 generates the unperturbed time development of the system:

$$x(t+dt) \doteq x(t) + dt \cdot f[x(t), p(t)], \quad t \in [t_0, t_1] \quad \cdot \quad \cdot \quad (2)$$

Time varying manipulation of the parameter set may be represented in its most general form (Tomović and Vukobratović, 1972) as

$$p^*(\epsilon, t) = p(t) + \epsilon v(t),$$
$$\epsilon \in R, \quad v(t) \in R^m, \quad t \in [t_k \geqq t_0, t_1] \quad \cdot \quad (3)$$

The variational component $v(t)$ of the perturbation $\epsilon v(t)$ is an arbitrary function in the parameter space R^m. The constant ϵ provides magnitude scaling. The resultant perturbation dynamics of the system is produced by solution of

$$\dot{x}^*(\epsilon, t) = f[x^*(\epsilon, t), p^*(\epsilon, t)],$$
$$x(t_0) = x_0, \quad x^*(\epsilon, t_k) = x(t_k), \quad t \in [t_0, t_1] \quad \cdot \quad (4)$$

This dynamics can be depicted as addition of the

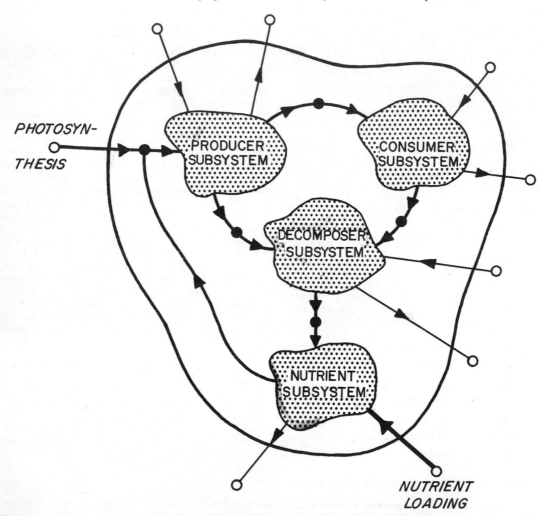

Figure 1. Generalized ecosystem, with material loop driven by energy, or energy flow constrained by matter recycling.

84

perturbation component to the nominal time development of the system (Astor, Patten, and Estberg, 1974):

$$x(t+dt) \doteq x(t) + dt \cdot f[x(t), p(t)] + \epsilon \cdot s(t),$$

$$t \in [t_0, t_1] \quad \ldots \ldots \ldots \quad (5)$$

Here,

$$s(t) = \lim_{\epsilon \to 0} \frac{x^*(\epsilon, t) - x(t)}{\epsilon}, \quad \ldots \ldots \quad (6)$$

$$s(t) \in R^n$$

is the time varying sensitivity of the system, computed to within a linear approximation by solution of

$$\dot{s}(t) \doteq \left[\frac{\partial f_i}{\partial x_j} \right] s(t) + \left[\frac{\partial f_i}{\partial p_k} \right] v(t), \quad \ldots \ldots \quad (7)$$

$$s(t_k \geq t_0) = 0, \quad t \in [t_0, t_1],$$

$$i, j = 1, \ldots, n, \quad k = 1, \ldots, m$$

For small perturbations ϵ, and for small time intervals $[t_0, t_1]$, $s(t)$ provides a measure of the deviation of states $x^*(\epsilon, t)$ from nominal states $x(t)$ induced by the parameter manipulations $\epsilon v(t)$. Solution of Equation 7 gives both the magnitude and timing of consequences of control actions, and as such provides a means of evaluating alternative management proposals. The entire parameter set $p(t)$ of the ecosystem model [Equation 1] is available for manipulation.

Reciprocal Dependence

The Jacobian matrices $[\partial f_i / \partial x_j]$ and $[\partial f_i / \partial p_k]$ of Equation 7 express the role of connectivity in signal propagation through the system's interactive network. The matrix $[\partial f_i / \partial x_j]$ defines the interdependence of state variables, and $[\partial f_i / \partial p_k]$ relates how parameter perturbations are distributed to the state variables.

At the solution (integral) level, the flow of cause in the system is seen in terms of the chain rule for implicit functions (Patten, 1969; Brylinski, 1972). That is, if x_i is influenced indirectly by p_k through a single dependency sequence (i.e., a cascade of state variables; Patten and Cale, 1974), then the sensitivity $s_i(t; p_k)$ of x_i to p_k is

$$s_i(t; p_k) = \left[\frac{\partial x_i}{\partial x_j} \cdots \frac{\partial x_\ell}{\partial x_k} \frac{\partial x_k}{\partial p_k} \right] \Delta p_k(t)$$

$$= \epsilon \left[\frac{\partial x_i}{\partial x_j} \cdots \frac{\partial x_\ell}{\partial x_k} \frac{\partial x_k}{\partial p_k} \right] v_k(t),$$

$$t \in [t_0, t_1] \quad \ldots \ldots \ldots \quad (8)$$

If there are multiple dependency cascades in parallel from p_k to x_i, then

$$s_i(t; p_k) = \epsilon \left[\sum_{\substack{\text{all} \\ \text{paths}}} \frac{\partial x_i}{\partial x_j} \cdots \frac{\partial x_\ell}{\partial x_k} \frac{\partial x_k}{\partial p_k} \right] v_k(t), \quad \ldots \quad (9)$$

$$t \in [t_0, t_1]$$

Finally, if there are feedback loops,

$$\frac{\partial x_a}{\partial x_b} \cdots \frac{\partial x_c}{\partial x_d} \frac{\partial x_d}{\partial x_a}$$

in the system, then Mason's rule applies. That is, letting

$$P_q = \frac{\partial x_i}{\partial x_j} \cdots \frac{\partial x_\ell}{\partial x_k} \frac{\partial x_k}{\partial p_k}$$

be the qth forward path from p_k to x_i, and letting

$$L_r = \frac{\partial x_a}{\partial x_b} \cdots \frac{\partial x_c}{\partial x_d} \frac{\partial x_d}{\partial x_a}$$

be the rth loop in the system interconnection, then the sensitivity of x_i to p_k is generated according to

$$s_i(t, p_k) = \epsilon \left[\frac{\sum_q P_q \prod_r (1 - L_r)}{\prod_r (1 - L_r)} \right] v_k(t), \quad \ldots \quad (10)$$

$$t \in [t_0, t_1]$$

In this expression it is understood that products of loops which touch one another are deleted to form the denominator, and in addition any denominator terms containing loops that touch path P_q are deleted to form the numerator loop expression (see, e.g., Huggins and Entwisle, 1968, for details). The point of Equation 10 is that state variables are influenced by parameter manipulations $\epsilon v(t)$ through complex patterns of feedforward and feedback dependency sequences. Each sequence (path or loop) is a chain of partial derivatives connecting the system states to the parameters $p(t)$, whose dynamics generate original cause (Patten and Cale, 1974). Each partial derivative $\partial x_i / \partial x_j$ expresses how a given state variable x_i is directly influenced by another x_j.

Ecologists have long debated the role of higher trophic levels in ecosystem function. One view is that consumers, particularly top consumers, are unimportant because they command only small amounts of the total energy and matter flowed and stored in ecosystems. Such a view probably contributes to the fact that eutrophication models tend to ignore consumers. Another view is that organisms at the tops of food chains perform a regulatory function. Ecologists who hold to this would partition the biota of ecosystems into process components (producers and decomposers through which most of the energy and material passes) and control components (consumers) which regulate the system irrespective of the small amounts of matter and energy involved. Such a partition is akin to the plant/controller division characteristic of cybernetic control systems. It is clear in ecosystems that parasites, pathogens, and man all exist on the energetic and material fringes, relatively immune to the dynamics of change at lower trophic levels, but capable, nevertheless, of exerting profound controlling influences. The basis for such power must lie not in the quantities of energy or matter flowing to or tied up in these organisms, but rather in their position in the interactive structure of ecosystems.

In particular, the following reciprocity principle may provide the amplification necessary for minority groups of organisms to exert great amounts of control. Referring to the previous discussion culminating in Equation 10, it may be observed that the dependency relationship between every pair of state variables in a system is always a reciprocal one. That is,

$$\frac{\partial x_i(t)}{\partial x_j(t)} = \frac{1}{\partial x_j(t)/\partial x_i(t)},$$

$$t \in [t_0, t_1], \; i,j = 1, \ldots, n \quad \ldots \ldots \ldots (11)$$

Since parameter influences are transmitted to states through chains of such partial derivatives [Equation 10], the contribution of one state variable in a causal chain leading to another is always the reciprocal of the influence of the latter in dependency sequences leading to the former.

Thus, if x_i (representing, say, a top consumer) changes only a small amount for a unit change in x_j (a primary producer), then the contribution of $\partial x_i/\partial x_j$ to dependency sequences leading from the parameter set $p(t)$ to the state variable set will be small, but the contribution of $\partial x_j/\partial x_i$ to sequences from $p(t)$ to $x(t)$ in which it appears will be correspondingly large. If $\partial x_i/\partial x_j = 10^{-6}$, for example, then $\partial x_j/\partial x_i = 10^6$, implying considerable amplification for a control capability even though x_i may represent an energetically and materially insignificant organism. The reciprocal effect may of course be masked in the entire dependency sequence, depending on the exact nature of the system interconnection [e.g., Equation 10]. But also, it may not be, and it could be argued that a strategy in the evolution of ecosystem structures is to achieve designs that permit expression of the control potential implicit in Equation 11.

On this point, the pyramid effect in ecosystems is germane. Trophic efficiencies are such that each higher trophic level contains at least an order of magnitude less energy and matter than the one preceding it. Thus, in dependency sequences going forward from the parameters of primary production, partial derivatives $\partial x_i/\partial x_j$,

where i refers to the next trophic level above j, tend to be less than unity. As when multiplying probabilities, multiplying numbers less than one in a dependency cascade without loops rapidly diminishes the influence of the originating parameter at each successive trophic level. When loops are present, the effect is further magnified because the denominator of Equation 10 tends to become larger and the whole expression smaller. Hence, the general rule would seem to be attenuation of influence along feedforward paths from primary production parameters, accounting for the well known buffering of upper trophic level organisms from producer dynamics. The return situation around the recycling loop is the opposite. Here, numbers $\partial x_j/\partial x_i$, where j is a lower trophic level than i, are generally greater than 10 (again, by the pyramid effect as from top consumers to decomposers). Multiplication of such numbers quickly produces amplification of influences of upper trophic level parameters and, as before, loops enhance the tendency. Thus, ecosystems in fact seem to be constructed to take advantage of the reciprocal relation [Equation 11], endowing upper trophic level organisms with significant inherent control capabilities.

These observations have great theoretical and practical potential. Theoretically, remote control may be more important in ecosystems than direct control. That is, parameters far removed from a state variable in the interconnection network may be more significant in controlling that state than other, more closely associated parameters. Sensitivity analysis of simple ecosystem models verifies this possibility (e.g., Patten, 1969; Brylinski, 1972; Waide et al., 1974). In applications, the reciprocal principle in combination with the knowledge that ecosystems are completely controllable can form the basis for broader avenues of attack on environmental problems.

In eutrophication research, the distinct possibility exists that small (inexpensive), well chosen and correctly timed (both determined from sensitivity analysis or optimal control theory) manipulations of ecosystem parameters which are remote from the nutrient-primary production subsystem may be able to produce dramatic improvements in water quality for the amount of control effort expended. Before the general principles enunciated above can be applied with assurance, however, the scope of "eutrophication" as a concept will have to be broadened to encompass entire lake or watershed ecosystems, and modeling of the eutrophication process will have to be extended to the other subsystems of Figure 1 that have so far been neglected.

References

Astor, P.H., B.C. Patten, and G.N. Estberg. 1974. The sensitivity substructure of ecosystems. *In* Systems Analysis and Simulation in Ecology, Vol. 4 (B.C. Patten, ed.). Academic Press. (In press.)

Brylinski, M. 1972. Steady-state sensitivity analysis of energy flow in a marine ecosystem. *In* Systems Analysis and Simulation in Ecology, Vol. 2 (B.C. Patten, ed.). Academic Press. New York. p. 81-101.

Huggins, W.H., and D.R. Entwisle. 1968. Introductory systems and design. Blaisdell. Waltham, Massachusetts.

Hutchinson, G.E. 1948. Circular causal systems in ecology. Ann. N.Y. Acad. Sci. 50:221-246.

Patten, B.C. 1969. Ecological systems analysis and fisheries science. Trans. Amer. Fish. Soc. 98:570-581.

Patten, B.C., and W.G. Cale. 1974. Propagation of cause in ecosystems. *In* Systems Analysis and Simulation in Ecology, Vol. 4 (B.C. Patten, ed.). Academic Press. (In press.)

Tomović, R., and M. Vukobratović. 1972. General sensitivity theory. Elsevier. New York.

Waide, J.B., J.E. Krebs, S.P. Clarkson, and E.M. Setzler. 1974. A linear systems analysis of the calcium cycle in a forested watershed ecosystem. *In* Progress in Theoretical Biology, Vol. 3 (R. Rosen, ed.). Academic Press. (In press.)

Zadeh, L.A., and C.A. Desoer. 1963. Linear system theory. The state space approach. McGraw-Hill, New York.

87

MULTI-NUTRIENT DYNAMIC MODELS OF ALGAL GROWTH

AND SPECIES COMPETITION IN EUTROPHIC LAKES[1]

V. J. Bierman, Jr., F. H. Verhoff, T. L. Poulson, and M. W. Tenney[2]

Introduction

Great concern has been expressed regarding the present and future quality of our freshwater lakes and mathematical modeling techniques can provide a systematic basis for a research approach and can greatly aid in the comparison of management options. People have begun to focus on various legal and technological control strategies for the purpose of reversing the deteriorated conditions in many of these lakes. In order to establish priorities for control and/or restoration, consideration must first be given to delineating the sources of pollution and evaluating their relative effects on the ecological system of the lake in question. The Environmental Protection Agency (1971b) has strongly endorsed the utilization of predictive mathematical modeling procedures in order to define allowable waste loads to receiving waters. Such models are to form the basis for the follow-up cost and/or optimization analysis required to meet basin water quality management plan objectives. Since the phenomenon of eutrophication is extremely complex, the development of a model detailing all of the biogeochemical interactions within a lake would be an extremely difficult task. Of necessity then, many judgments must be made in constructing such models.

It is the purpose of this paper to focus on kinetic equations describing algal growth in eutrophic environments. Recently, increasing attention has been given to the role of cell physiology in the occurrence of intense algal blooms, especially those cellular processes involving phosphorus and nitrogen. For example, luxury nutrient uptake and possible multi-nutrient growth regulation can greatly influence the development of an algal crop. In this study, a new model of microbial growth is further developed and is applied to the case of algal growth in the surface waters of eutrophic freshwater lakes. Simulations involving one- and two-species systems are presented for the case where phosphorus is the primary regulating nutrient. In addition, a more general two-species system is investigated for the case where both phosphorus and nitrogen are important regulating nutrients. For purposes of simplicity, growth simulations are extended only through the course of a single growing season and do not include the effects of fall overturns that would generally occur in stratified lakes. However, temperature variation, cell sinking, cell decomposition, nutrient recycle, and predation by higher animals are taken into account.

In short, the basic purpose of the model presented in the paper is to give *insight* and not to produce exact numerical values.

Pertinent Literature

Generally, the classic equation by Monod (1949) or some other similar empirical function is used as the growth kinetics component in overall ecosystems models as well as in the analysis of laboratory growth data (Chen, 1970; DiToro et al., 1970; Middlebrooks et al., 1969; Thomann et al., 1970; Gaudy et al., 1971). The basis for the development of the Monod equation was the observation that the specific growth rate and the concentration of the assumed growth limiting nutrient could be related by a hyperbolic function under certain laboratory conditions. However, it will be shown that this function cannot model the well-known and extremely important phenomenon of luxury nutrient uptake, nor is it clear that the generalization of the Monod formalism to the multi-limiting nutrient case has any broad application or validity. Other authors have developed different empirical expressions relating cell growth rate to limiting substrate concentration; for example, Teissier's (1936) exponential expression for growth rate, and the two-phase proposal of Garret and Sawyer (1952). Kono (1968) developed a model for the case of batch cultivation taking into account four phases of growth. He based his model on a similarity to chemical kinetics and introduced the new concepts of

[1]This research was supported in part by a pre-doctoral fellowship from the National Wildlife Federation to Mr. Bierman and by a research and development grant (R-801245) from the Environmental Protection Agency (Water Quality Office) entitled "Eutrophic Lake Reclamation by Physical and Chemical Manipulations." The results presented herein constitute part of Mr. Bierman's doctoral dissertation for the Graduate School, University of Notre Dame.

[2]V. J. Bierman, Jr. is with the Department of Civil Engineering and F. H. Verhoff is with the Department of Chemical Engineering, University of Notre Dame, Notre Dame, Indiana; T. L. Poulson is with the Department of Biological Sciences, University of Illinois at Chicago Circle, Chicago, Illinois; and M. W. Tenney is President, TenEch Environmental Consultant, South Bend, Indiana.

critical concentration and the coefficient of consumption activity. Powell et al. (1967) modified Monod's equation to include the effects of mass transfer. He compared his growth model with that of Monod and with that of Teissier and found that his model fitted the experimental data better than either of these. The important point to note about these models is that they are primarily empirical, that is, the experimental data were fit to a functional form. For example, Monod used the curve, found for the rate of enzymatic reaction, to relate growth rate to limiting substrate concentration. However, none of the authors propose a mechanism for growth because it is always assumed that the metabolic processes are quite complex involving a variety of rate laws, and that these processes are among the rate determining steps for growth.

Grenney et al. (1973) have developed a three-compartment mathematical model to describe phyto-plankton growth in a nitrogen-limited environment and studied species competition as a function of intracellular nitrogen storage. They parameterized their model in terms of the intracellular concentrations of inorganic nitrogen, intermediate nitrogenous organics, and protein. These chemical species were then related by functions of the Monod-Michaelis-Menten form. The fact that the metabolic process within the cell is quite complex and still not completely understood is well known, but the assumption that knowledge about these rates and their dependency on substrate and enzyme concentrations is necessary for a mechanistic determination of limiting-substrate growth rate dependence is questionable.

In the writers opinion, only a few facts about the metabolic activity of the cell are required along with a knowledge of the mode of growth in order to explain the relationship between growth rate and limiting substrate. An analogous example from engineering which illustrates the importance of the mode of operation is the problem of producing a solution with a specified concentration of B from a solution containing a certain concentration of A. Although the kinetics of the reaction of A to give B may be fully known, it is necessary to know the mode of operation before the relationship between initial concentration of A and final concentration of B is known. To know the mode of operation, one needs to know if the reaction is to be carried out in a batch reactor, stirred tank reactor, a plug flow reactor, or any combination of these. Also, the residence time or time of reaction in each of these reactors must be known. In fact, it is often possible to predict the relationship between output of concentration B to the input of concentration A with only a sketchy knowledge of the kinetics and a thorough knowledge of the mode of operation. It is this basic approach which underlines the development of the growth model presented in this paper.

Mechanism of Microbial Growth

Preliminary versions of the growth model and some of its components, along with extensive applications, have already been presented in a series of papers (Verhoff et al., 1972a, 1972b, 1971, 1973; Cordeiro and Verhoff, 1972). Conceptually, cell growth was treated as a two-step mechanism which involves separate nutrient uptake and cell synthesis processes and includes an intermediate nutrient storage capability. The mode of cell growth, that is, assimilation and subsequent ingestion of nutrients, is judged to be more important in modeling growth dynamics than are the extremely complex chemical conversions occurring within the cell.

For purposes of simplicity, consider the case of a single algal species for which phosphorus is the primary growth regulating factor. Equation 1 is a symbolic representation of the growth model for this case.

$$A + P \xrightarrow{R1} B \xrightarrow{R2} (1+x)A \quad \ldots \ldots \quad (1)$$

In words, a viable, phosphorus-deficient algal cell, A, combines with an amount of phosphorus, P, at the net reaction rate, R1, to form an ingesting cell, B. The ingesting cell then forms $(1 + x)$ units of A at the rate R2, which is directly dependent on the internal level of phosphorus in cell B. This process may or may not include actual cell division. The value of x is determined by a myriad of possible factors, including the phosphorus level in cell B, the minimum possible phosphorus content of cell A, the value of R2 and the time interval over which R2 proceeds, and the rate of cell lysis. It is well-known that algal cells have the ability to accumulate and store both phosphorus and nitrogen to levels which are significantly in excess of their immediate metabolic needs (Kuhl, 1967; Gerloff and Skoog, 1954). Recently, there has been increasing evidence that algal growth rates are most directly determined by the internal cell levels of these nutrients for cases where either one of them is the primary growth regulating factor (Fuhs, 1969; Fuhs et al., 1971; Azad and Borchardt, 1970; Rhee 1972; Kholy, 1956; Caperon, 1968; Eppley and Thomas, 1969).

Much of the writers' work has involved the development of a reaction-diffusion transport mechanism to model the initial nutrient uptake step described by the rate R1 in Equation 1. A detailed presentation appears elsewhere (Verhoff et al., 1973) and only the specific application to the present case will be considered here. The following mechanism by which cell A can accumulate phosphorus is suggested by various experimental observations:

$$T_1 + S_1 \rightleftharpoons T_2 \quad \ldots \ldots \ldots \ldots \quad (2)$$

$$T_2 + S_2 \rightleftharpoons T_3 \quad \ldots \ldots \ldots \ldots \quad (3)$$

in which

S_1	=	substrate 1, phosphate
S_2	=	substrate 2, a cation (potassium or sodium)
T_1, T_2, T_3	=	substrate-carrier complexes

Essentially, the carrier T_1 is postulated to combine with phosphate at the cell surface to form the complex T_2 by means of a reversible reaction. The complex T_2 then combines reversibly with the cation S_2 and the resulting complex T_3 diffuses to the interior of the cell where it can disassociate back into the free phosphate and cation. The carrier T_1 can then cycle back to the cell surface and pick up additional phosphate. The mechanism given by Equations 2 and 3 is consistent with evidence that phosphate uptake by microorganisms is a carrier-mediated process which transports this nutrient against its own concentration gradient (Harold et al., 1965; Harold and Baarda, 1968b; White and MacLeod, 1971; White et al., 1968; Medveczky and Rosenberg, 1971; Carberry and Tenney, 1972). The cation co-transport feature of this mechanism is consistent with additional evidence indicating a strong correlation between cation uptake, primarily potassium, and phosphate uptake (Scott, 1945; Schmidt et al., 1949; Goodman and Rothstein, 1957). It is quite possible that this correlation involves direct interaction of potassium with the phosphate transport system (Weiden et al., 1967; Harold and Baarda, 1968a; Carberry and Tenney, 1972; Medveczky and Rosenberg, 1971).

Now the application of the transport formalism to Equations 2 and 3 yields the following expression for the net rate of phosphorus transport:

$$J_1 = K \left[\frac{1}{1 + K_1 C_{1L} + K_1 K_2 C_{1L} C_{2L}} - \frac{1}{1 + K_1 C_{10} + K_1 K_2 C_{10} C_{20}} \right] \quad \cdot \cdot (4)$$

in which

J_1	=	net phosphorus transport rate
C_{10}, C_{20}	=	external medium concentrations of phosphorus and potassium, respectively
C_{1L}, C_{2L}	=	internal cell concentrations of phosphate and potassium, respectively
K_1, K_2	=	equilibrium constants for the reactions in Equations 2 and 3, respectively
K	=	constant to be determined

The expression in brackets in Equation 4 is essentially a dimensionless "weighting factor" which reflects the balance between the phosphorus nutrition of the cell and the availability of phosphorus in the external medium at any given time. *This feedback capability is the central feature of the transport mechanism.* Assuming that phosphorus is available, transport into the cell will be maximum when the cell is most "hungry" and will approach zero as the cell approaches complete satiation. In fact, phosphorus can even "leak" back into the medium under certain conditions since the expression in brackets can be either positive or negative. The enhancement of phosphate uptake by cation co-transport can best be illustrated by assuming that a cell has become fully loaded with phosphorus and that J_1 is now equal to zero. In this case, Equation 4 can be rearranged in the following manner:

$$\frac{C_{1L}}{C_{10}} = \frac{1 + K_2 C_{20}}{1 + K_2 C_{2L}} \cdot \cdot \cdot \cdot \cdot \cdot \cdot \cdot \cdot (5)$$

Thus the maximum ratio of internal phosphorus concentration to external phosphorus concentration becomes inversely proportional to this same ratio for the potassium concentrations. According to this mechanism, an increase in the external potassium concentration will, within limits, allow the cell to accumulate a greater amount of phosphorus.

A detailed application of the transport mechanism in Equations 2 and 3 to the case of glycine and sodium co-transport in pigeon red cells has been presented by Verhoff (1972). Unfortunately, insufficient data exist at this time for a corresponding application to the case of phosphorus transport in algal cells. For the applications in this paper, the terms in Equation 4 which refer to cation concentrations will be considered negligible. If this is done, the actual transport mechanism becomes that given only by Equation 2 and the net phosphorus transport rate becomes

$$J_1 = K \left[\frac{1}{1 + K_1 C_{1L}} - \frac{1}{1 + K_1 C_{10}} \right] \quad \cdot \cdot \cdot \cdot (6)$$

The essential feature of carrier-mediation and feedback capability is still retained in this working formulation. For purposes of compatibility with notation to be introduced for other parameters in the proposed growth model and for the case of nitrogen-dependent growth, Equation 6 will be re-written in the following manner:

$$R1SP = R1MP \left[\frac{1}{1 + (PK1)(PCA)} - \frac{1}{1 + (PK1)(PCM)} \right]$$
$$\cdot \cdot \cdot \cdot \cdot \cdot \cdot (7)$$

in which

R1SP	=	*specific phosphorus* uptake rate in step 1 of the growth model in (time)$^{-1}$
R1MP	=	*maximum phosphorus* uptake rate in step 1 of the growth model in (time)$^{-1}$
PK1	=	equilibrium constant for reaction between phosphorus and carrier in (liters/mole)
PCA	=	*phosphorus concentration* inside *A* cells in (moles/liter)

PCM = *phosphorus concentration* in external *medium* in (moles/liter)

Physically, PK1 represents the degree of affinity between the external phosphorus and the phosphorus carrier. The value of PK1 determines the threshold value of PCM for effective net transport into the cell and is an important factor in determining the rate of the transport, R1SP.

Once R1SP has been determined, it is necessary to calculate the amount of phosphorus removed from the medium in a given interval of time. This necessarily depends upon the concentration of the algal crop that is present:

$$R1P = (R1SP)(A) \ldots \ldots \ldots (8)$$

in which

R1P = phosphorus removal from solution in (mg/liter•time)

A = algae in (mg dry wt/liter)

NOTE:

$$R1SP = \frac{mg\ phosphorus}{(mg\ algae)(time)}$$

$$R1P = \frac{mg\ phosphorus}{(mg\ algae)(time)} \times \frac{mg\ algae}{liter}$$

$$R1P = \frac{mg\ phosphorus}{(liter)(time)}$$

The differential in the medium phosphorus concentration, DPCM, during a transport time interval, TTRAN, is given by

$$DPCM = (R1P)(TTRAN) \ldots \ldots \ldots (9)$$

Now in order to relate this change in concentration to the mass of new algal cells to be produced in the ingestion step, the phosphorus removed from solution must be distributed over the mass of the algae present; thus (after converting DPCM to moles/liter)

$$DPSA = DPCM/A \ldots \ldots \ldots \ldots (10)$$

in which

DPSA = *differential* of *phosphorus* storage in *A* cells in (mole•P/mg algae)

Once the phosphorus is inside the cells, the model transforms most of it into intermediate storage compounds and retains only a small fraction of it as a soluble inorganic pool. Thus,

$$PSA = PSA + \left[\frac{CONP-1}{CONP}\right](DPSA) \ldots \ldots (11)$$

$$PCA = PCA + \left[\frac{1}{CONP}\right](DPSA)(DDEN) \ldots (12)$$

in which

PSA = *phosphorus storage* in *A* cells in (moles•P/mg algae)

PCA = *phosphorus concentration* in *A* cells in (moles•P/liter)

DDEN = cell *dry weight-density* factor in (mg dry weight/liter cell volume)

CONP = *concentration factor* for *phosphorus*

Luxury phosphorus uptake can involve concentration factors on the order of 10^5 to 10^6 (Provasoli, 1969). On the basis of various chemical fraction techniques, rapid conversion of assimilated inorganic phosphate to various intermediate phosphorus compounds has been observed in a number of different microorganisms, including algae (Rhee, 1972; Miyachi et al., 1964; Schmidt et al., 1946; Goodman and Rothstein, 1957; Harold et al., 1965; Harold, 1962; Weiden et al., 1967; and Medveczky and Rosenberg, 1971). This is a complex set of phenomena and the different fractionation techniques which exist will not always identify the same intermediate phosphorus compounds for a given case. Since the writers contend that such detailed information is not necessary in order to develop a mechanistic model of growth dynamics, the formalism adopted in Equations 11 and 12 represents a convenient approach to the problem. In this way, the model is parameterized only in terms of quantities that can be easily measured. The concentration factor, CONP, can be calculated in a straightforward manner using cell size, maximum total phosphorus storage, and phosphorus concentration in the external medium (Kuenzler and Ketchum, 1962). In effect, CONP is a "driving potential" for phosphorus accumulation.

The ingestion step begins with the calculation of the actual total amount of phosphorus per cell:

$$P = (PSB)(FACT) \ldots \ldots \ldots \ldots (13)$$

in which

P = total moles•P/cell

PSB = total *phosphorus storage* in *B* cells in (moles•P/mg algae)

FACT = conversion factor (mg dry weight/cell)

Mass growth rate is then related to P using the following equation by Fuhs (1969) and Fuhs et al. (1971):

$$R2P = R2PT\left[1 - EXP\left[-.693(P/PO-1)\right]\right] \ldots (14)$$

in which

R2P = phosphorus-dependent growth rate expressed as fractional increase in mass per day; i.e., $(day)^{-1}$

R2PT = maximum possible temperature-dependent growth rate in $(day)^{-1}$

PO = minimum cell content of P in (moles P•/cell)

Maximum growth rate is achieved with this equation when P becomes approximately six times PO. When P exceeds this so called critical concentration, luxury storage of phosphorus results and the growth rate remains constant at its maximum value. Cell growth will cease when P reaches its minimum possible level, PO. Fuhs based his conclusions on work using two marine diatoms and several species of bacteria under phosphorus-limited conditions. His results are generally consistent with the data of Rhee (1972) and Azad and Borchardt (1970) who also studied algal growth as a function of internal phosphorus levels. Soeder et al. (1971) have recently pointed out that the minimal P content per cell is a species-specific figure of limited statistical variation and they emphasized the importance of this quantity in estimating maximum possible algal crops for natural waters of known phosphorus budgets, rather than using optimal or average cell P levels. The differential increase in algal crop, DA, during the synthesis time interval, TSYN, is calculated as

$$DA = (R2PT)(A)(TSYN) \quad \ldots \ldots \ldots (15)$$

The total crop is then adjusted by setting

$$A = A + DA$$

Cell sinking, cell decomposition, nutrient recycle, and grazing by higher animals are all handled by additional straightforward calculations which, for reasons of space, will not be presented here. The following equations are used to model the effect of temperature on growth rates:

$$T = TMAX\left[.50 + .50 \left[SIN\left(6.28 \cdot DAY/360\right)\right]\right].(16)$$

$$R2PT = R2MP(T/TOPT) \quad \ldots \ldots \ldots (17)$$

in which

T	=	actual temperature of surface waters on given day
TMAX	=	maximum temperature of surface waters during growing season
R2PT	=	maximum possible growth rate on *given day* for optimum phosphorus level
R2MP	=	maximum possible growth rate for optimum phosphorus level *and* optimum temperature
TOPT	=	optimum temperature for growth for given algal species

Equation 16 was chosen on the basis of a reasonably accurate fit to the data of Sweet (1969) for Stone Lake in Cassopolis, Michigan. For this purpose, Stone Lake is a typical, stratified, eutrophic lake. If temperature exceeds TOPT then R2PT is set equal to R2MP. The incorporation of the effect of light intensity depends on the particular application. The approach adopted here is to assume complete mixing in the epilimnion and to select R2MP as a function of the ratio of "illuminated depth" to "stirred depth" (Mortimer, 1969). This ratio is critical in determining the amount of time that an algal cell will spend in a zone of optimum illumination for growth. In this way,

R2MP represents an effective maximum growth rate for the entire stirred depth or epilimnion.

This completes the essential outline of the growth model for the case of a single algal species where phosphorus is the primary growth regulating nutrient. Applications to this as well as to more complicated systems will follow.

Application to a Single-Species Phosphorus-Limited System

Methods

The first example will involve the development of the common nuisance-bloom forming blue-green alga, *Microcystis aeruginosa,* as a function of the soluble phosphorus concentration in the epilimnion of a typical eutrophic lake beginning immediately after spring overturn. The values for the physiological parameters used in this simulation are:

R2MP	= .25 day^{-1}
PO	= .58x10^{-15} moles P/cell
FACT	= .25x10^{-7} mg dry wt/cell
R1MP	= .50x10^{-2} hr^{-1}
PK1	= .50x10^{7} liters/mole
CONP	= .50x10^{6}
DDEN	= .20x10^{6} mg dry wt/liter cell volume
TMAX	= 25°C
TOPT	= 35°C
TTRAN	= 1 hr
TSYN	= .5 day

The value for R2MP is intuitively reasonable and the resulting algal growth curve matches well the timing of the blue-green blooms that have been observed in Stone Lake (DePinto, unpublished). PO and FACT can be easily calculated from the data of Gerloff et al. (1952), and the "Algal Assay Procedure-Bottle Test" (1971), respectively. The Joint Industry/Government Task Force on Eutrophication (1969) strongly emphasized the importance of ultimately expressing algal concentration as a dry weight equivalent, no matter what actual primary assay procedure is used. R1MP was calculated on the basis of the general phenomenon that P-starved algae, when placed in a medium containing excess phosphorus, can absorb up to approximately thirty times their minimum cell P content in about half a day (Kuenzler and Ketchum, 1962; Azad and Borchardt, 1970; Ketchum, 1939b; Kholy, 1956). This approach also makes sense intuitively because phosphorus "eating" rate then becomes proportional to the phosphorus requirement of the particular species. The value of PK1 is difficult to determine directly because most kinetics data are interpreted and expressed in terms of the Monod-Michaelis-Menten growth formalism. PK1 is a parameter describing only the transport system which is not directly related to actual cell growth, a distinction which has also been pointed out by Shapiro et al. (1969). However, especially in the case of blue-green algae, it is not difficult to infer a realistic value for PK1. The range of soluble phosphorus concentrations for most natural waters is approximately 10-50 µ g•P/l (Brown et al., 1971). Sawyer (1944), in his classic paper, indicated that summer nuisance-blooms of blue-green algae could be expected to occur when soluble phosphorus exceeded 10 µg•P/l and combined nitrogen exceeded 300 µg•N/l at spring overturn. Gentile and Maloney (1969) and Stewart et al. (1970) observed in the laboratory that two different P-starved, blue-green algae were both quite responsive to phosphorus concentrations less than 10 µ g•P/l. Blue-green algae thus seem to operate well at the lower extreme of the range of naturally-occurring phosphorus levels. For starved cells, that is, as PCA approaches zero in Equation 7, the above-indicated value for PK1 corresponds to a half-maximum rate of phosphorus uptake when the concentration in the external medium is approximately 6 µg•P/l. The value for CONP is consistent with those reported by Provasoli (1969). The value of DDEN used here is the same as that used by Kuenzler and Ketchum

(1962). The values for TMAX and TOPT correspond to those given by Sweet (1969) and Fogg (1950), respectively. Finally, for purposes of simplicity, this first example will not incorporate nutrient recycle and an arbitrary rate of cell lysing at 3 percent per day will be applied to the standing crop. Most of the simulations presented in this paper were performed on a General Automation Model 18/30 computer with 16K of core space. A simple linear interpolation method was used and the indicated values of TTRAN and TSYN gave more than adequate resolution.

Results

Figure 1 shows the lag effect between the decrease in the external phosphorus concentration and the increase in algal growth. The shaded portion of the algal growth curve (solid line) represents the algal development after 50 percent of the phosphorus (dotted line) has been removed from solution. Almost 85 percent of the maximum crop develops after this point. This lag effect is a direct consequence of the mode of cell growth and luxury

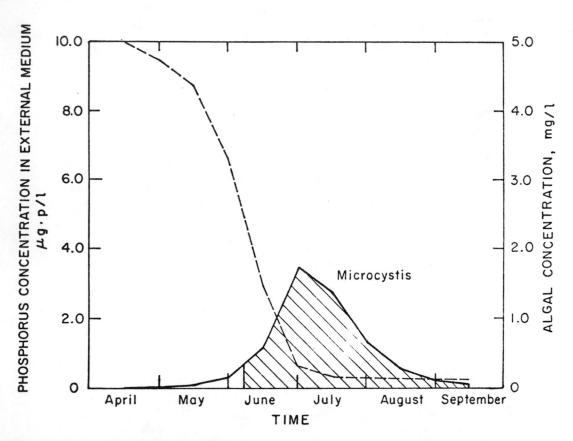

Figure 1. There is a lag between the decrease in the external phosphorus concentration (dotted line) and the increase in algal growth (solid line) for the single-species case.

phosphorus uptake. The net growth rate actually continues to increase beyond this same point. This is caused by the combined effect of luxury phosphorus uptake and steadily increasing temperature. The peak crop is observed at a point in time when the external phosphorus concentration is less than $1 \mu g \cdot P/l$. This is generally considered to be below the level of detectability in routine lab analyses. Such an inverse relationship between algal crop and phosphorus concentration is actually observed in many lakes (e.g., Schindler et al., 1971). Figure 2 shows the time course of development of the internal phosphorus level in the cells. The indicated minimum cell level corresponds to the value PO, the level at which growth rate is equal to zero. The indicated critical concentration is that value of P from Equation 14 at which growth rate becomes approximately maximum. Recall that this critical concentration is about six times PO. The condition of luxury phosphorus uptake is maintained essentially from mid-April through the third week in June. During this period, cell growth rate is independent not only of the external phosphorus concentration but also of the internal phosphorus level. This is an actual phenomenon (Shapiro, 1970) and it is an important consideration in attempting to relate algal crops to observe nutrient concentrations in lakes. When the internal phosphorus is exhausted, growth rate decreases rapidly and the algal bloom "crashes" because there is essentially no more available phosphorus in the medium to refuel the bloom.

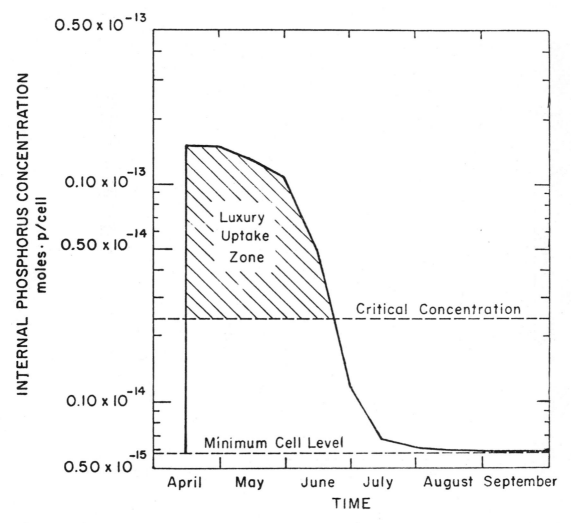

Figure 2. The internal cell phosphorus concentration as a function of time shows luxury phosphorus uptake for the single-species case.

95

Figure 3 illustrates the incompatibility between the growth rates predicted by the proposed model and the growth rates predicted using conventional Monod kinetics for the example presented. The solid line represents the time course of development of the ratio specific growth rate to maximum growth rate for each given day of the example. It can be seen, upon comparison with Figure 2, that constant maximum growth rate is actually obtained as long as P is greater than the critical internal concentration. This is true even though external phosphorus decreases steadily from mid-April onward. Specific growth rate does not decrease until P falls below its critical concentration late in June. Growth effectively ceases as P approaches PO and only the basic, structural, phosphorus is left in the cells. The growth equation of Monod, expressed in terms of the notation used in this paper, is given by

$$R2P = R2PT \left[\frac{PCM}{PCM + K_p} \right] \quad \ldots \ldots (18)$$

in which

K_p = value of PCM for which specific growth rate, R2P, is equal to R2PT/2; that is, at half-maximum growth rate

It is recognized that a direct comparison between the two growth models is not possible because of their different mechanisms; however, it can be shown that it is impossible to select *any* value for K_p such that the predictions of the two models become equivalent. The dotted lines in Figure 3 also represent the ratio specific growth rate to maximum growth rate, but for Equation 18 instead of the proposed model. The values used for PCM are those shown in Figure 1 as functions of time. Now for $K_p \doteq 6$ $\mu g \cdot P/l$, Equation 18 does not achieve maximum growth rate and it does not predict the lag in the fall-off of the specific growth rate due to luxury uptake. If $K_p = 25$ $\mu g \cdot P/l$, both of these negative features become even more obvious. If K_p is now decreased to 1 $\mu g \cdot P/l$, Equation 18 gives a better fit to the specific growth rate and the discrepancy from the lag effect is not as severe. However, this value of K_p is already too low because a non-physical

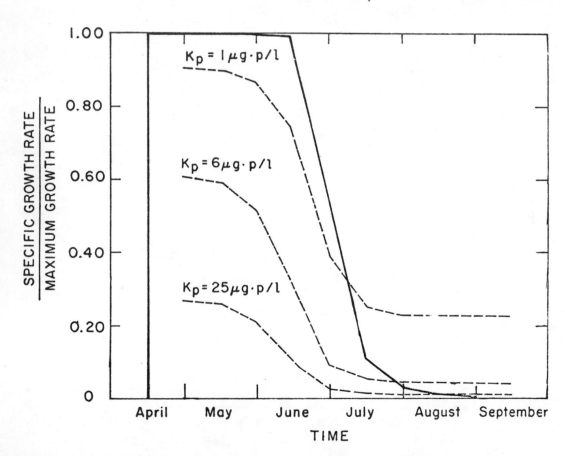

Figure 3. The specific growth rate predictions of the proposed model (solid line) and of Monod kinetics (dotted lines) are incompatible for the single-species case.

solution is now obtained for the time period after mid-July. Equation 18 gets "stuck" at approximately 20 percent of maximum growth rate during this period; that is, for lower growth rates, the external phosphorus concentration must be lower during this period. A unique value cannot be found for K_p that will enable the Monod equation to fit the growth rates predicted by the proposed two-step model at both high and low phosphorus concentrations.

To summarize, the primary difference between the proposed two-step growth model and the usual Monod approach is the ability of the proposed mechanism to model the lag effect between phosphorus removal from solution and the resulting algal growth. Algal yield becomes a variable whose value is determined by the physiological condition of the cells at any given point in time.

Application to a Two-species Phosphorus-limited System

Method

This example will involve competition between a typical green alga, *Chlorella,* and the blue-green alga *Microcystis* for a limited supply of phosphorus. The investigation of algal species competition can be a very important and fruitful area of research. Nutrient-deficient lakes usually have small standing crops and high algal species diversity, while culturally eutrophic lakes generally have large algal crops and a few dominant algal species (Fruh, 1967). Most often these dominant species consist of blue-green algae which tend to cause the most difficult nuisance problems (Edmondson, 1966). Thus it becomes important to know if a given set of environmental conditions will tend to favor a stable or an unstable system. One important difference between blue-greens and the more desirable green algae is the degree to which each species is grazed by high animals (Richerson, 1971; Lund, 1969). In addition, there are important physiological differences between these two groups which can greatly affect their relative growth characteristics.

The values for the physiological parameters of the *Microcystis* will be the same as those used in the previous example. The corresponding values for the *Chlorella* will be the following:

R2MP =	.35 day^{-1}
PO =	.32x10^{-14} moles•P/cell
FACT =	.27x10^{-7} mg dry wt/cell
R1MP =	.20x10^{-1} hr^{-1}
PK1 =	.10x10^{7} liters/mole
CONP =	.10x10^{6}
TMAX =	25°C
TOPT =	25°C

The choice of .35 day^{-1} as the maximum growth rate for *Chlorella* was based on the proportion between the maximum growth rates for *Chlorella* and *Microcystis*

under laboratory conditions that most closely resemble natural environmental conditions. This must be carefully elaborated because some confusion does exist on this point. The development of blue-green algae is often mentioned as being correlated with warm waters (e.g., Vinyard, 1966) and attempts have been made to discern possible causative relationships (e.g., Hammer, 1964). However, the actual situation is complex and any such relationships remain unclear (Lund, 1965). It has been reported that the optimum temperature for growth of most blue-green algae, excluding thermophilic strains, is around 35°C (Fogg, 1950; Castenholz, 1966). Kratz and Myers (1955) found that a special, high-temperature strain of *Anacystis nidulans* had a growth optimum of 41°C. In contrast, most green algae, including *Chlorella,* achieve optimum growth rate at around 25°C (Marre', 1962). The crucial point is that the surface waters of most temperate-zone lakes of any significant size do not generally exceed 25-26°C (Moyle, 1968; Megard, 1970; Sweet, 1969). Another distinction that must be made is the one between photosynthesis and actual cell growth. Not only can these be separate processes, but Sorokin (1959) has shown that their temperature optima can be quite different in the same alga.

It was in recognition of the many different sets of procedures and test conditions that have been employed to study algal growth that the Joint Industry/Government Task Force on Eutrophication (1969) developed the Provisional Algal Assay Procedure. One of the primary goals of this JTF was to devise a test procedure, the results of which could be applied, with judgment, to field conditions. Two of the important features of this procedure are that a constant temperature of 24°C is used for all algal species tested, and that atmospheric CO_2 is used as the primary source of carbon. Under these conditions it has been observed that the optimum growth rate of *Selenastrum,* the standard green test alga, is approximately double the optimum growth rates of both *Microcystis* and *Anabaena,* the two standard blue-green algae (EPA, 1971a). This observation is also consistent with similar data by Payne (personal communication) who used the subsequently revised version of the P.A.A.P. nutrient medium, the so-called Normal Algal Assay Medium (Weiss and Helms, 1971). The growth rate of *Selenastrum* is comparable to the optimum growth rates of other typical green algae such as *Chlorella* and *Scenedesmus* (Azad and Borchardt, 1970). Concerning carbon, Morton et al. (1971), in contrast to many algal growth studies which use CO_2-enriched air, investigated the growth of *Chlorella, Microcystis,* and *Anabaena* at dissolved CO_2 concentrations that were in equilibrium with or were lower than atmospheric CO_2 concentrations. Although the concentrations of nitrogen and phosphorus in the nutrient media used by Morton were not at all representative of environmental levels, it was still observed that *Chlorella* grew substantially faster than either of the two blue-green algae. Accordingly, for purposes of the example to be presented, the maximum growth rate of *Chlorella* was chosen to be approximately double that of *Microcystis, at 25°C* (see Equation 17), and TMAX and TOPT were both set equal to 25°C.

The values for PO and FACT can be easily calculated from the data of Barber (1968) and Kholy (1956) by using the previously indicated dry weight-density factor, DDEN. The maximum phosphorus eating rate, R1MP, was calculated in the same manner as the corresponding rate for *Microcystis*. The phosphorus requirement for *Chlorella* is considerably higher than that for *Microcystis*.

For starved cells (i.e., as PCA approaches zero in Equation 7), the indicated value for PK1 corresponds to a half-maximum rate of phosphorus uptake when the concentration in the external medium is approximately 30 $\mu g \bullet P/l$. Recall that the corresponding concentration for *Microcystis* is only 6 $\mu g \bullet P/l$. Physically, this means that *Microcystis* is much more efficient at phosphorus uptake than is *Chlorella* at low external concentrations. Soeder et al. (1971) and Uhlmann (1971) have cited evidence indicating that green algae need relatively high phosphorus concentrations for optimal growth. It appears that *Chlorophyta* prefer levels of at least 20-40 $\mu g \circ P/l$. Shapiro (1973) has reported that the half-saturation constants for phosphate uptake by blue-green algae are significantly lower than those for green algae over a wide range of environmental conditions. The consequences of this phenomenon are consistent with field observations. It is well-known that green algae are the dominant species in nutrient-rich sewage lagoons (e.g., Porges and Mackenthun, 1963) and Bush and Welch (1972) have observed that blue-green algae, especially *Microcystis*, can apparently out-compete other species for phosphorus when external concentrations are low. Hammer (1964) has also reported the high efficiency of *Microcystis* for phosphorus utilization. Finally, the indicated value of CONP is consistent with the data of Azad and Borchardt (1970) and with Provasoli (1969). The values of CONP for *Chlorella* and *Microcystis* have been made proportional to the individual values of PK1 for these algae.

In order to make the simulations for the two-species systems as realistic as possible, cell sinking, cell decomposition, nutrient recycle and predation by higher animals will be included. This will be done for the case where phosphorus is the primary regulating nutrient and also for the case where both phosphorus and nitrogen are important regulators. Both algal species in each case will be lysed at the rate of .03 day^{-1}, based on total crop. Luxury-stored phosphorus and/or nitrogen will be immediately released from these cells in the surface waters (Foree et al., 1970) and the resulting dead algal material will be bacterially decomposed at the rate of .03 day^{-1}, based on total non-refractory mass (Jewell and McCarty, 1971). Green algae will be considered to sink out of the epilimnion and to be grazed by higher animals at rates both equal to .03 day^{-1}, based on total crop. Blue-green

algae will be sunk at .015 day^{-1} and will not be subject to grazing (Richerson, 1971). Where not otherwise indicated, the rate constants for the above processes are based on intuitively reasonable values.

Results

Figure 4 shows that *Microcystis*, in spite of its slower growth rate, will easily dominate *Chlorella* at low external phosphorus concentrations (dotted line). This is a direct consequence of the much greater efficiency of the phosphorus transport system in the *Microcystis*. At 10 $\mu g \bullet P/l$, the transport system in the *Microcystis* is more than half-saturated, while the same system in *Chlorella* is not even close to half-saturation. The maximum *Microcystis* crop in Figure 4 is significantly higher than the maximum crop in Figure 1. This is due to nutrient recycle, which was not included in the first example. With recycle, one can obtain a greater fraction of the total possible theoretical crop for a fixed amount of limiting nutrient. This is because the nutrients released upon cell lysis and decomposition can be re-synthesized into new algal cells instead of remaining unused. Figure 5 shows that the maximum *Microcystis* crop has increased almost proportionately when the initial external phosphorus concentration was increased from 10 $\mu g \bullet P/l$ to 25 $\mu g \bullet P/l$. *Chlorella* still does not develop significantly, in spite of the fact that its phosphorus transport system is almost half-saturated at the beginning of the simulation. The primary reasons for this are luxury phosphorus uptake by the *Microcystis* and the much greater phosphorus requirement of the *Chlorella* in comparison to the *Microcystis*. The transport system of the *Microcystis* remains well-saturated at the phosphorus concentrations present in solution (dotted line) for approximately the first two months and considerable luxury uptake occurs during this period. The average degree of saturation of the phosphorus transport system in *Chlorella* is much less than half during this same period and this alga will not be able to accumulate much phosphorus over and above its immediate metabolic needs. This inability to compete as well for the limited supply of phosphorus is further accentuated by the fact that the minimum cell phosphorus level of *Chlorella* is approximately five times that of *Microcystis*. In addition, *Chlorella* is subject to a higher sinking rate and grazing by higher animals.

Figure 6 gives an indication of the inherent stability that can be obtained in a two-species system as opposed to a single-species system. The maximum crop of *Microcystis* has increased only slightly, in spite of the fact that the initial phosphorus concentration was doubled from 25 $\mu g \bullet P/l$ to 50 $\mu g \bullet P/l$. Almost all of the additional phosphorus is utilized by the *Chlorella*. For most of the first six weeks of the simulation, the phosphorus transport system of the *Chlorella* is more than half-saturated. It is only under such conditions that this alga can exploit its relatively fast eating and fast growing rates. After six weeks, the *Chlorella* cannot replenish its internal phosphorus fast enough to maintain a net positive growth rate against the negative effects of sinking, lysing, and grazing. It is these effects, together with its high phosphorus requirement, that keep the maximum crop of *Chlorella* low compared to the maximum *Microcystis* crop. In effect, the *Chlorella* acts as an important "buffer" on the size of the subsequent summer blue-green bloom.

In summary, all three of the plots for this present application have shown the frequently-observed correlation between nuisance blooms of blue-green algae and low nutrient concentrations, and also the familiar positive correlation of these blooms with warm summer waters. The central point of this application is that species differences in phosphorus uptake efficiencies and in minimum phosphorus requirements are *sufficient* causes to explain these phenomena and it is not *necessary* to invoke a causal relationship between elevated temperatures and assumed higher growth rates for blue-green algae as compared to other species. In fact, it should be stressed again that the maximum possible growth rate for *Microcystis* in this application is only one-half of the maximum possible growth rate for *Chlorella* at the highest tempera-

ture that actually occurs during the entire summer. Also, the possible increase in stability of a two-species algal system, as compared to a single-species system, is suggested. It will be shown in the following example that this stability can even be enhanced by the additional constraint of a second regulating nutrient.

Application to a Two-species System Limited by Both Phosphorus and Nitrogen

Methods

This example will again involve competition between *Chlorella* and *Microcystis*, but for limited supplies of both phosphorus and nitrogen. The general approach for nitrogen regulation of growth will be the same as that used for the case of phosphorus as outlined in Equations 1 through 13. For each transport and synthesis interval, parallel and independent calculations are made for both phosphorus and nitrogen. For each synthesis interval of 0.5 day, separate specific growth rates are calculated based on the internal cell levels of these two nutrients.

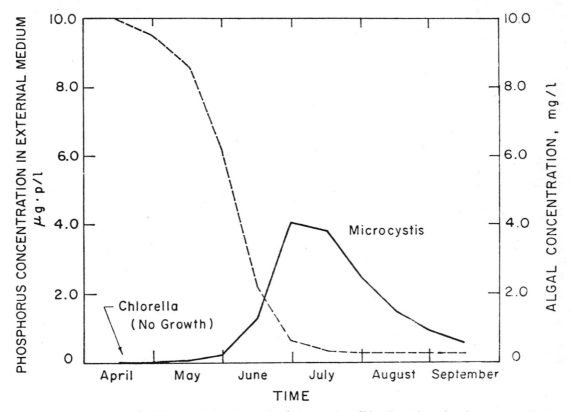

Figure 4. The slower-growing *Microcystis* dominates the faster-growing *Chlorella* at low phosphorus concentrations (dotted line).

The lower, or controlling, of these two rates is then selected as the *actual* specific growth rate to be used during this interval (Equation 15). This entire process is carried out independently for each species so that it is possible for one of them to be phosphorus-regulated and the other to be nitrogen-regulated during the same synthesis interval.

Although results are not presented in this paper, it is also possible to investigate interactions between phosphorus and nitrogen in the transport step using the general version of the transport formalism that we have developed (Verhoff et al., 1973). Evidence consistent with such interactions has been reported by Ketchum (1939a). For example, a nitrate ion might compete with the similarly-charged monobasic phosphate ion for the same enzyme carrier. This competition might also involve the co-transport of a cation to maintain electroneutrality.

The following equation, similar to Equation 7, is used for calculating nitrogen transport rate:

$$R1SN = R1MN \left[\frac{1}{1 + (NK1)(NCA)} - \frac{1}{1 + (NK1)(NCM)} \right]$$

$$\cdots \cdots \cdots (19)$$

in which

R1SN =	*specific nitrogen* uptake rate in step 1 of growth model in (time)$^{-1}$	
R1MN =	*maximum nitrogen* uptake rate in step 1 of the growth model in (time)$^{-1}$	
NK1 =	equilibrium constant for reaction between nitrogen and carrier in (liters/mole)	
NCA =	*nitrogen concentration* inside *A* cells in (moles/liter)	
NCM =	*nitrogen concentration* in external *medium* in (moles/liter)	

The parameters CONN and R2MN represent, respectively, nitrogen concentration factor and the maximum possible growth rate for optimum nitrogen level and temperature. Nitrogen-dependent growth rate is related to internal cell

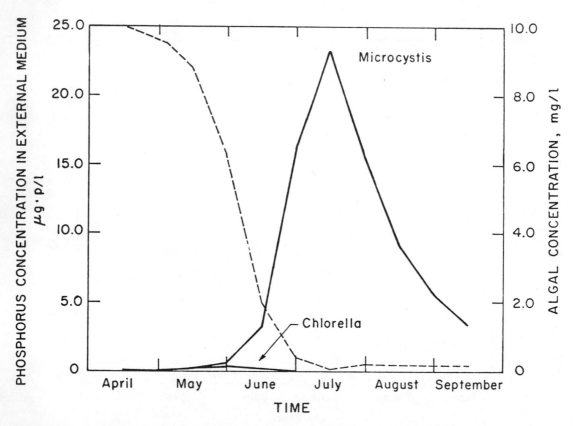

Figure 5. *Microcystis* **increases proportionately after phosphorus concentration (dotted line) has been increased from 10** μg •P/l to 25 μg° P/l.

100

nitrogen level using the following equation based on work by Caperon (1968):

$$R2N = R2NT \left[\frac{(N-NO)}{HALFN + (N-NO)} \right] \quad \dots (20)$$

in which

R2N = nitrogen-dependent growth rate expressed as fractional increase in mass per day; i.e., $(day)^{-1}$

R2NT = maximum possible temperature-dependent growth rate in $(day)^{-1}$

NO = minimum cell content of nitrogen in (moles•N/cell)

N = total moles•N/cell

HALFN = value of (N-NO) for which growth rate is half-maximum.

The quantity (N-NO) represents the amount of cell nitrogen in excess of the minimum level. When N equals NO then cell growth rate becomes zero. R2NT is related to temperature in the same manner as for phosphorus shown in Equation 17.

The same numerical values for the growth parameters used in the previous example for *Chlorella* and *Microcystis* will be used in this present example. The additional parameters introduced by the consideration of nitrogen will be the following:

MICROCYSTIS

NO   = $.522 \times 10^{-13}$ moles•N/cell
R1MN = $.30 \times 10^{-2}$ hr^{-1}
NK1   = $.10 \times 10^7$ liters/mole
CONN = $.10 \times 10^7$

CHLORELLA

NO   = $.52 \times 10^{-13}$ moles•N/cell
R1MN = $.30 \times 10^{-2}$ hr^{-1}
NK1   = $.10 \times 10^7$ liters/mole
CONN = $.10 \times 10^7$

The value of NO for *Microcystis* is based on data by Gerloff et al. (1952). For *Chlorella*, NO can be calculated

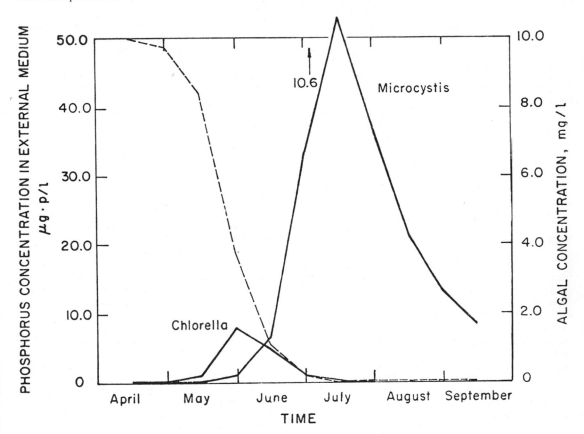

Figure 6. A degree of inherent stability can be obtained in a two-species system as opposed to a single-species system because *Microcystis* does not respond proportionately in this case after phosphorus concentration (dotted line) has been increased from 25 µg•P/l to 50 µg•P/l.

101

using the parameter FACT and the data of Fitzgerald (1969). The value for R1MN used for both algae is consistent with the data of Syrett (1956) and Thomas and Krauss (1955). Fitzgerald (1968) has shown that there is no significant difference in the nitrogen (NH_4^+) uptake rates for *Chlorella* and *Microcystis*. It was on this basis that NK1 was also set equal for both algae at a value of $.10x10^7$. For starved cells (i.e., as NCA approaches zero in Equation 19), this corresponds to a half-maximum rate of nitrogen uptake when the concentration in the external medium is approximately 15 μ g∘N/l. This is generally consistent with the data of Eppley and Thomas (1969) and Carpenter and Guillard (1971) for various species of marine phytoplankton. Also, Pearson et al. (1969) reported that the green alga *Selenastrum* had a Monod constant for growth of approximately 15 μ g• N/l. The value for CONN is reasonable and is based on typically observed ranges for luxury nitrogen uptake (e.g. Fitzgerald, 1969). The value for HALFN can be determined by relating N and NO at half-maximum cell growth rate. Based on Caperon's (1968) results, HALFN was set equal to $.102x10^{-13}$ for the example presented.

Note that, unlike for phosphorus, the minimum cell levels of nitrogen for *Chlorella* and *Microcystis* are very similar. The internal molar ratios N/P, expressed in terms of minimum cell levels, are 16 for *Chlorella* and 90 for *Microcystis*. Average values for this ratio are generally taken to be approximately 15 (e.g. Pearson et al., 1969). Thus *Microcystis,* in relation to phosphorus, appears to have quite a high nitrogen requirement. The available data, at least for *Chlorella* and *Microcystis,* seem to indicate that species differences in nitrogen uptake efficiencies and in nitrogen requirements may not be as significant in lakes as are the corresponding differences for phosphorus. Perhaps a more important mechanism of algal competition for nitrogen is the ability of some blue-green algae to fix atmospheric nitrogen instead of competing for dissolved combined nitrogen.

Results

In Figure 7, the development of the *Microcystis* is limited by available nitrogen when the initial nutrient concentrations are 10 μg• P/l and 60 μ g• N/l, corresponding to an N/P molar ratio of 13. Contrast this with the much larger *Microcystis* crop in Figure 4 for the identical *Chlorella-Microcystis* system, but without nitrogen dependence. In Figure 8, the *Microcystis* is no longer nitrogen-limited because the initial nitrogen concentration has been raised to 300 μ g•N/l for the same initial phosphorus concentration of 10 μ g• P/l. The molar ratio N/P in this case is equal to 66. Because of low initial phosphorus concentration, the *Chlorella* still does not develop significantly in the present example.

The above N/P ratios are based on actual field data. The case where N/P is equal to 66 corresponds to the nitrogen and phosphorus values suggested by Sawyer (1944) as being conducive to nuisance blooms of blue-green algae. The value of N/P equal to 13, chosen as a convenient fraction of Sawyer's ratio, is consistent with data tabulated by Moyle (1968) for 45 Minnesota Lakes identified as having summer nuisance blooms.

In effect, the simulations for only phosphorus-dependent growth assume that an infinite supply of nitrogen is available. In these cases, all of the available phosphorus is actually synthesized into cellular material and algal growth will not cease until internal phosphorus reaches its minimum level in each given species. If nitrogen dependence is added, algal growth will cease when either internal nitrogen *or* internal phosphorus reaches the minimum level in each species. Of course, this is not completely true for the case of nitrogen-fixing algal species. For the simulation in Figure 8, there is sufficient nitrogen present for the *Microcystis* to use virtually all of the phosphorus in the system. For the case in Figure 7, nitrogen deficiency causes a much smaller *Microcystis* crop because not all of the phosphorus in the system can be used.

In Figures 9 and 10, at equal initial phosphorus concentrations of 50 μ g•P/l, different N/P ratios not only cause significant changes in maximum crops, but also changes in species composition. In Figure 9, the initial nutrient concentrations are 50 μ g•P/l and 300 μg•N/l, corresponding to an N/P ratio of 13. In Figure 10, the concentrations are 50 μ g•P/l and 1500 μ g•N/l, corresponding to an N/P ratio of 66. Again, contrast these two figures with Figure 6 showing the identical *Chlorella-Microcystis* system, but without nitrogen dependence. Only for the case in Figure 10 is there enough nitrogen so that the algae can utilize virtually all of the phosphorus in the system.

Chlorella is the dominant species in Figure 9 because the initial phosphorus concentration is sufficiently high that it can exploit its relatively faster growth rate and utilize most of the limited supply of nitrogen before the *Microcystis.* In Figure 10 there is enough nitrogen in the system for both algal species to utilize the available phosphorus, virtually to the best of their individual capabilities. In this case, *Microcystis* is dominant for the same reasons that it dominated in the case of Figure 6.

In summary, it is shown that the availability of nitrogen can greatly affect the size and species composition of algal crops for a given amount of phosphorus. Whenever algal growth is modeled as a function of only one nutrient, an implicit assumption is made that all other nutrients necessary for growth are present in virtually infinite amounts and can never affect growth rates, standing crops, or species composition. In dealing with such a complex phenomenon as eutrophication it is doubtful that there exists a unique nutrient for which this assumption is valid. The results presented in this section, however, can only be applied with great discretion because the potentially very important effects of nitrogen-fixing algae have not been included.

Work is continuing on the development of a three-species model which does include a nitrogen-fixer and preliminary results have been obtained.

Summary and Conclusion

This paper has been focused primarily on kinetic equations describing algal growth in the surface waters of eutrophic lakes. A two-step growth model, involving separate nutrient transport and cell synthesis mechanisms, was further developed and was applied to one- and two-species systems where phosphorus was the primary nutrient regulating growth and to a two-species system where both phosphorus and nitrogen were important regulators. Growth simulations were extended through the course of a single growing season and effects of temperature, cell sinking, cell decomposition, nutrient recycle, and predation by higher animals were included. The basic purpose of the model was to give *insight* and not to produce exact numerical values.

For the case of phosphorus-dependent growth of a single algal species, *Microcystis,* the proposed model predicts a lag between the removal of phosphorus from solution and the subsequent algal growth. Because of luxury phosphorus uptake, 85 percent of the algal crop developed after 50 percent of the phosphorus had been removed from solution. The specific growth rate predictions of Monod kinetics were not compatible with those of the proposed model for the example presented.

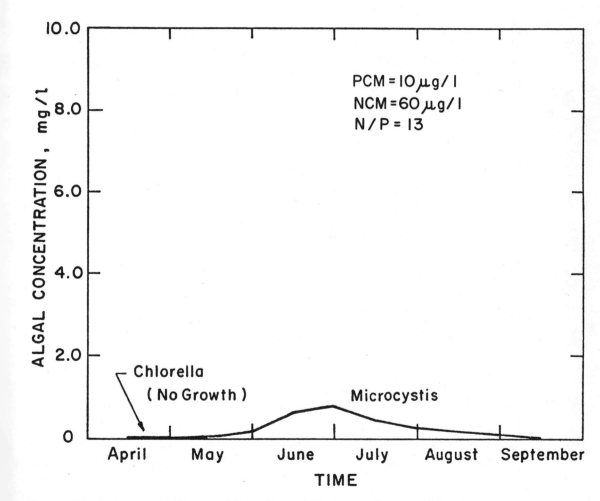

Figure 7. The development of *Microcystis* is limited by available nitrogen when the initial nutrient concentrations are 10 μg⋅P/l and 60 μg⋅N/l, corresponding to an N/P ratio of 13.

The first two-species application involved the competition of *Chlorella* and *Microcystis* for a fixed amount of phosphorus. Because of a greater phosphorus transport efficiency, the slower-growing *Microcystis* dominated the faster-growing *Chlorella* at an initial phosphorus concentration of 10 μ g• P/l. When the initial phosphorus concentration was increased from 10 μ g• P/l to 25 μ g•P/l, the maximum crop of *Microcystis* increased almost proportionately, indicating that most of the additional phosphorus was taken up only by this alga. When the initial phosphorus concentration was increased from 25 μ g•P/l to 50 μ g•P/l, most of the additional phosphorus was now taken up by the *Chlorella* and the *Microcystis* crop increased only slightly because the *Chlorella* was better able to exploit its faster growth rate at this higher phosphorus concentration. In effect, the *Chlorella* acted

as a "buffer" on the size of the subsequent summer *Microcystis* bloom.

The second two-species application involved the competition of *Chlorella* and *Microcystis* for fixed amounts of both phosphorus and nitrogen. We have shown that the availability of nitrogen, in relation to a given amount of phosphorus, can greatly affect total algal crop as well as species composition. For initial nutrient concentrations of 10 μ g• P/l and 60 μ g•M/l, the growth of *Microcystis* was N-limited and it could not utilize all of the available phosphorus. When initial nitrogen concentration was increased to 300 μ g• N/l for this case, the *Microcystis* crop was much larger because this alga was now able to utilize almost all of the available phosphorus. *Chlorella* did not develop significantly at the low initial

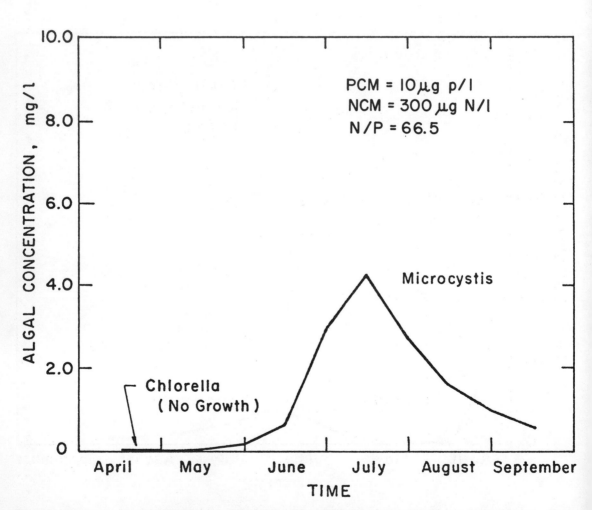

Figure 8. *Microcystis* uses virtually all of the available phosphorus when the initial nutrient concentrations are 10 μ g• P/l and 300 μg• N/l, corresponding to an N/P ratio of 66.

104

phosphorus concentration of 10 μ g•P/l. For initial nutrient concentrations of 50 μ g•P/l and 300 μ g•N/l, *Chlorella* dominated *Microcystis* because, at this high phosphorus concentration, *Chlorella* was able to exploit its faster growth rate and it used up most of the nitrogen before the *Microcystis* could develop sufficiently. When initial nitrogen was increased to 1500 μ g•N/l for this case, both algae were able to utilize the available phosphorus to the best of their individual capabilities and *Chlorella* dominated early in the growing season and *Microcystis* dominated in mid-summer. We recognized that the results of this application must be carefully qualified because we did not include the effects of nitrogen-fixation.

From this study, we conclude that it is not necessary to have a detailed knowledge of the complex metabolic processes that occur inside microorganisms in order to develop a realistic, analytic model of algal growth.

From the results of the two-species system in which phosphorus is the regulating nutrient, we conclude that species differences in phosphorus uptake efficiencies and in minimum phosphorus requirements are *sufficient* causes to explain the frequently observed correlation between nuisance blooms of blue-green algae and low nutrient concentrations, and also the familiar positive correlation of these blooms with warm summer waters. It is not *necessary* to invoke a causal relationship between elevated temperatures and assumed higher growth rates for blue-green algae as compared to other species. Also, it appears that a higher degree of internal stability can be

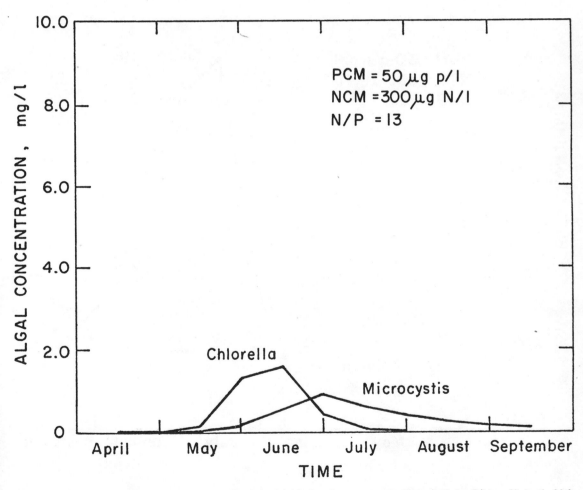

Figure 9. *Chlorella* **dominates** *Microcystis* **because the initial phosphorus concentration of 50** μg• P/l is sufficiently high that it can exploit its growth rate advantage and utilize most of the limited supply of nitrogen, 300 μ g• N/l, before the *Microcystis*.

obtained in a two-species system as opposed to a single-species system.

Since the introduction of nitrogen dependence can significantly affect total algal crop as well as species composition, we believe that such dependence is an important part of any algal growth model which is to be applied to field conditions. This dependence should also include the potentially very important effects of nitrogen fixation.

Work is continuing on the effects of nitrogen fixation and nutrient inputs into the lake.

References

Azad, H.S., and J. S. Borchardt. 1970. Variations in phosphorus uptake by algae. Environmental Science and Technology 4:737-743.

Barber, J. 1968. The influx of potassium into *Chlorella pyrenoidosa*. Biochimica et Biophysica Acta 163:141-149.

Brown, R. L., D. B. Porcella, and D. Toerien. 1971. Phosphorus and eutrophication. *In* Proceedings of Seminar on Eutrophication and Biostimulation, Clear Lake, California, California Department of Water Resources, October 19-21.

Bush, R. M., and E. B. Welch. 1972. Plankton associations and related factors in a hypereutrophic lake. Water, Air, and Soil Pollution 1:257-274.

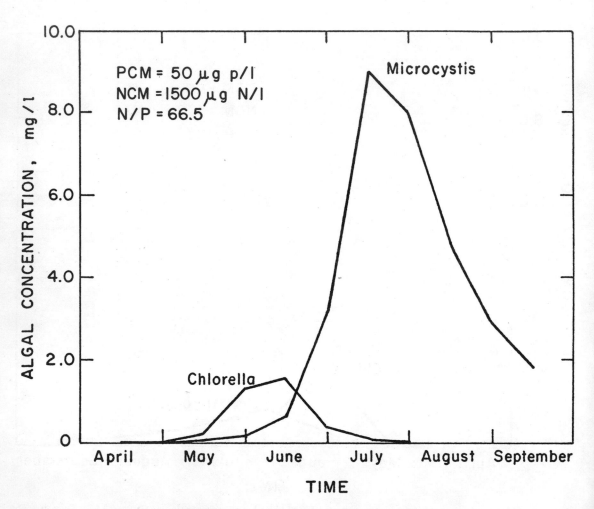

Figure 10. There is enough nitrogen in the system, 1500 μg•N/l, for both *Chlorella* and *Microcystis* to utilize the initial phosphorus concentration of 50 μg•P/l to the best of their capabilities.

Caperon, J. 1968. Population growth response of *Isochrysis galbana* to nitrate variation at limiting concentrations. Ecology 49:866.

Carberry, J. B., and M. W. Tenney. 1972. The mechanism of luxury uptake of phosphate by activated sludge. Paper presented at 45th Annual Conference, Water Pollution Control Federation, Atlanta, Georgia, October 8-13.

Carpenter, E. J., and R. R. L. Guillard. 1971. Intraspecific differences in nitrate half-saturation constants for three species of marine phytoplankton. Ecology 52:183-185.

Castenholz, R. W. 1966. Environmental requirements of thermophilic blue-green algae. *In* Environmental Requirements of Blue-Green Algae, Proceedings of a Symposium, Corvallis, Oregon, September 23-24.

Chen, C. W. 1970. Concepts and utilities of ecological model. Journal of the Sanitary Engineering Division, American Society of Civil Engineers 96(SA5):1085-1097.

Cordeiro, C. F., and F. H. Verhoff. 1972. A mathematical model of the surface microbial populations of a freshwater lake. Paper presented at the 65th Annual Meeting of the American Institute of Chemical Engineers, New York, November.

DePinto, J. V. No date. Department of Civil Engineering, University of Notre Dame, unpublished results.

DiToro, D. M., D. J. O'Connor, and R. V. Thomann. 1970. A dynamic model of phytoplankton populations in natural waters. Environmental Engineering and Science Program, Manhattan College, New York City, June.

Edmondson, W. T. 1966. Why study blue-green algae? *In* Environmental Requirements of Blue-Green Algae, Proceedings of Symposium, Corvallis, Oregon, September 23-24.

Environmental Protection Agency. 1971a. Algal assay procedure-bottle test. National Eutrophication Research Program, August.

Environmental Protection Agency. 1971b. Guidelines - water quality management planning. Water Quality Office, Washington, D.C., January.

Eppley, R. W., and W. H. Thomas. 1969. Comparison of half-saturation constants for growth and nitrate uptake of marine phytoplankton. Journal of Phycology 5:375.

Fitzgerald, G. P. 1968. Detection of limiting or surplus nitrogen in algae and aquatic weeds. Journal of Phycology 4:121-126.

Fitzgerald, G. P. 1969. Field and laboratory evaluations of bioassay for nitrogen and phosphorus with algae and aquatic weeds. Limnology and Oceanography 14:206-212.

Fogg, G. E. 1950. The comparative physiology and biochemistry of the blue-green algae. Bacteriological Reviews 20:148-165.

Foree, E. G., W. J. Jewell, and P. L. McCarty. 1970. The extent of nitrogen and phosphorus regeneration from decomposing algae. *In* Advances in Water Pollution Research, Proceedings of the 5th International Conference, San Francisco, Vol. 2.

Fruh, E. G. 1967. The overall picture of eutrophication. Journal of the Water Pollution Control Federation 39:1449-1463.

Fuhs, G. W. 1969. Phosphorus content and rate of growth in the diatoms *Cyclotella nana* and *Thalassiosira fluriatilis*. Journal of Phycology 5:312-321.

Fuhs, G. W., S. D. Demmerle, E. Canelli, and M. Chen. 1971. Characterization of phosphorus-limited plankton algae. *In* Nutrients and Eutrophcation: The Limiting Nutrient Controversy, Proceedings of a Symposium, American Society of Limnology and Oceanography, Michigan State University, East Lansing, Michigan, February 11-12.

Garrett, M. T., Jr., and C. N. Sawyer. 1952. Proceedings of 7th Industrial Waste Conference, Purdue.

Gaudy, A. F., Jr., A. Obayashi, and E. T. Gaudy. 1971. Control of growth rate by initial substrate concentration at values below maximum rate. Applied Microbiology 22:1041-1047.

Gentile, J. H., and T. E. Maloney. 1969. Toxicity and environmental requirements of a strain of *Aphanizomenon flos-aquae* (L.) Ralfs. Canadian Journal of Microbiology 15:165-173.

Gerloff, G. C., G. P. Fitzgerald, and F. Skoog. 1952. The mineral nutrient of *Microcystis aeruginosa*. American Journal of Botany 39:26-32.

Gerloff, G. C., and F. Skoog. 1954. Cell contents of nitrogen and phosphorus as a measure of their availability for growth of *Microcystis aeruginosa*. Ecology 35(3):348-353.

Goodman, J., and A. Rothstein. 1957. The active transport of phosphate into the yeast cell. Journal of General Physiology 40:915-923.

Grenney, W. J., D. A. Bella, and H. C. Curl, Jr. 1973. A theoretical approach to interspecific competition of phytoplankton communities. The American Naturalist 107:405-425.

Hammer, V. T. 1964. The succession of "bloom" species of blue-green algae and some causal factors. International Assocation of Theoretical and Applied Limnology 15:829-836.

Harold, F. M. 1962. Depletion and replenishment of the inorganic polyphosphate pool in *Neurospora crassa*. Journal of Bacteriology 83:1047-1057.

Harold, F. M., and J. R. Baarda. 1968a. Effects of nigericin and monactin on cation permeability of *Streptococcus faecalis* and metabolic capacities of potassium-depleted cells. Journal of Bacteriology 95:816.

Harold, F. M., and J. R. Baarda. 1968b. Inhibition of membrane transport in *Streptococcus faecalis* by uncouplers of oxidative phosphorylation and its relationship to proton conduction. Journal of Bacteriology 96:2025.

Harold, F. M., R. L. Harold, and A. Abrams. 1965. A mutant of *Streptococcus faecalis* defective in phosphate uptake. Journal of Biological Chemistry 240:3145-3153.

Jewell, W. J., and P. L. McCarty. 1971. Aerobic decomposition of algae. Environmental Science and Technology 5:1023-1031.

Joint Industry/Government Task Force on Eutrophication. 1969. Provisional algal assay procedure. P.O. Box 3011, Grand Central Station, New York, New York, February.

Ketchum, B. H. 1939a. The absorption of phosphate and nitrate by illuminated cultures of *Nitzschia closterium*. American Journal of Botany 26:399-406.

Ketchum, B. H. 1939b. The development and restoration of deficiencies in the phosphorus and nitrogen composition of unicellular plants. Journal of Cellular and Comparative Physiology 13:373-381.

Kholy, A. A. 1956. On the assimilation of phosphorus in *Chlorella pyrenoidosa*. Physiologia Plantarum 9:137-143.

Kono, T. 1968. Kinetics of microbial cell growth. Biotechnology and Bioengineering 10:105-131.

Kratz, W. A., and J. Myers. 1955. Nutrition and growth of several blue-green algae. American Journal of Botany 42:282-287.

Kuenzler, E. J., and B. H. Ketchum. 1962. Rate of phosphorus uptake by *Phaeodactylum tricornutum*. Biological Bulletin 123:134.

Kuhl, A. 1967. Phosphate metabolism of green algae. *In* Algae, Man and the Environment, D. F. Jackson, Ed., Proceedings of an International Symposium sponsored by Syracuse University and the New York State Science and Technology Foundation, Syracuse, New York, June 18-30.

Lund, J. W. G. 1965. The ecology of freshwater phytoplankton. Biological Reviews 40:231-293.

Lund, J. W. G. 1969. Phytoplankton. *In* Eutrophication: Causes, Consequences, Correctives-Proceedings of a Symposium, National Academy of Sciences, Washington, D.C., 306-330.

Marre', E. 1962. Temperature. *In* Physiology and Biochemistry of Algae, R. A. Lewin, Ed., Chapter 33, Academic Press.

Medveczky, N., and H. Rosenberg. 1971. Phosphate transport in *Escherichis coli*. Biochimica et Biophysica Acta 241:494.

Megard, R. O. 1970. Lake Minnetonka: Nutrients, nutrient abatement, and the photosynthetic system of the phytoplankton. Interim Report No. 7, Limnological Research Center, University of Minnesota, Minneapolis, December.

Middlebrooks, E. J., T. E. Maloney, C. F. Powers, and L. M. Kaack, Eds. 1969. Proceedings of the Eutrophication-Biostimulation Assessment Workshop, University of California, Berkeley, and Federal Water Pollution Control Administration, Northwest Laboratory, Corvallis, Oregon.

Miyachi, S., R. Kanai, S. Mihara, S. Miyachi, and S. Aoki. 1964. Metabolic roles of inorganic polyphosphate in *Chlorella* cells. Biochimica et Biophysica Acta 93:625-634.

Monod, J. 1949. The growth of bacterial culture. Annual Review of Microbiology 3:371.

Mortimer, C. H. 1969. Physical factors with bearing on eutrophication in lakes in general and in large lakes in particular. *In* Eutrophication: Causes, Consequences, Correctives-Proceedings of a Symposium, National Academy of Sciences, Washington, D.C., 340-368.

Morton, S. D., P. H. Derse, and R. C. Sernau. 1971. The carbon dioxide system and eutrophication. Office of Research and Monitoring, Environmental Protection Agency, November.

Moyle, J. B. 1968. Notes on some characteristics of Minnesota lakes having blue-green algal blooms. Special Publication No. 52, State of Minnesota, Department of Conservation, Division of Game and Fish, April.

Payne, A. G. No date. The Proctor and Gamble Co., Ivorydale Technical Center, Cincinnati, Ohio, personal communication.

Pearson, E. A., E. J. Middlebrooks, M. Tunzi, A. Adinarayana, P. H. McGauhey, and G. A. Rohlich. 1969. Kinetic assessment of algal growth. *In* Proceedings of the Eutrophication-Biostimulation Assessment Workshop, University of California, Berkeley, and Federal Water Pollution Control Administration, Northwest Laboratory, Corvallis, Oregon.

Porges, R., and K. M. Mackenthun. 1963. Waste stabilization ponds: Use, functions, and biota. Biotechnology and Bioengineering 5:255-273.

Powell, E. O., C. G. T. Evans, R. E. Strange, and D. W. Tempest, Eds. 1967. Microbial physiology and continuous cultures. H. M. Stationery Office, London, p. 34.

Provasoli, L. 1969. Algal nutrition and eutrophication. *In* Eutrophication: Causes, Consequences, Correctives-Proceedings of a Symposium, National Academy of Sciences, Washington, D.C., p. 574-593.

Rhee, G. 1972. Competition between an alga and an aquatic bacterium for phosphate. Limnology and Oceanography 17:505-514.

Richerson, P. J. 1971. The role of zooplankton in the process of eutrophication. *In* Proceedings of Seminar on Eutrophication and Biostimulation, Clear Lake, California, California Department of Water Resources, October 19-21.

Sawyer, C. N. 1944. Fertilization of lakes by agricultural and urban drainage. Journal of the New England Water Works Association 61:925.

Scott, G. T. 1945. The mineral composition of phosphate-deficient cells of *Chlorella pyrenoidosa* during the restoration of phosphate. Journal of Cellular and Comparative Physiology 26:35-42.

Shapiro, J. 1970. A statement on phosphorus. Journal of the Water Pollution Control Federation 42(5):Part 1.

Shapiro, J. 1973. Blue-green algae: Why they become dominant. Science 179:382-384.

Shapiro, J., W. Chamberlain, and J. Barrett. 1969. Factors influencing phosphate use by algae. Advances in Water Pollution Research, Proceedings of the 4th International Conference, Prague, p. 149-169.

Schindler, D. W., F. A. J. Armstrong, S. K. Holmgren, and G. J. Brunskill. 1971. Eutrophication of Lake 227, experimental lakes area, Northwestern Ontario, by addition of phosphate and nitrate. Journal Fisheries Research Board of Canada 28:1763-1782.

Schmidt, G., L. Hecht, and S. J. Thannhauser. 1946. The enzymatic formation and the accumulation of large amounts of a metaphosphate in bakers' yeast under certain conditions. Journal of Biological Chemistry 166:775.

Schmidt, G., L. Hecht, and S. J. Thannhauser. 1949. The effect of potassium ions on the absorption of orthophosphate and the formation of metaphosphate by bakers' yeast. Journal of Biological Chemistry 178:733-742.

Soeder, C. J., H. Muller, H. D. Payer, and H. Schuller, 1971. Mineral nutrition of planktonic algae; some considerations, some experiments. International Association of Theoretical and Applied Limnology 19:39-58.

Sorokin, C. 1959. Tabular comparative data for the low- and high-temperature strains of *Chlorella*. Nature 184:613-614.

Stewart, W. D. P., G. P. Fitzgerald, and R. H. Burris. 1970. Acetylene reduction assay for determination of phosphorus availability in Wisconsin lakes. Proceedings of the National Academy of Sciences, 66:1104-1111.

Sweet, M. A. 1969. M.S. Thesis, Department of Civil Engineering, University of Notre Dame.

Syrett, P. J. 1956. The assimilation of ammonia and nitrate by nitrogen-starved cells of *Chlorella vulgaris* II. The assimilation of large quantities of nitrogen. Physiologia Plantarum 9:19-37.

Teissier, G. 1936. Ann. Physiol. Physiochem. Biol., 12:527.

Thomann, R. V., D. J. O'Connor, and D. M. DiToro. 1970. Modeling of the nitrogen and algal cycles in estuaries. Paper presented at Fifth International Water Pollution Research Conference, July.

Thomas, W. H., and R. W. Krauss. 1955. Nitrogen metabolism in *Scenedesmus* as affected by environmental changes. Plant Physiology 30:113-122.

Uhlmann, D. 1971. Influence of dilution, sinking and grazing rate on phytoplankton populations of hyperfertilized ponds and micro-ecosystems. International Association of Theoretical and Applied Limnology 19:100-124.

Verhoff, F. H., J. B. Carberry, V. J. Bierman, Jr., and M. W. Tenney. 1973. Mass transport of metabolites, especially phosphate in cells. American Institute of Chemical Engineers, Symposium Series, 129, 69:227-240.

Verhoff, F. H., W. F. Echelberger, Jr., M. W. Tenney, P. C. Singer, and C. F. Cordeiro. 1971. Lake water quality prediction through systems modeling. Proceedings 1971 Summer Computer Simulation Conference, Boston, Mass., p. 1014-1023.

Verhoff, F. H., and K. R. Sundaresan. 1972. A theory of coupled transport in cells. Biochimica et Biophysica Acta 255:425.

Verhoff, F. H., K. R. Sundaresan, and M. W. Tenney. 1972. A mechanism of microbial growth. Biotechnology and Bioengineering 14:411.

Vinyard, W. C. 1966. Growth requirements of blue-green algae as deduced from their natural distribution. *In* Environmental Requirements of Blue-Green Algae, Proceedings of a Symposium, Corvallis, Oregon, September 23-24.

Weiden, P. L., W. Epstein, and G. Schultz. 1967. Cation transport in *Escherichia coli* VII. Potassium requirement for phosphate uptake. Journal of General Physiology 50:1641-1661.

Weiss, C. M., and R. W. Helms. 1971. The inter-laboratory precision test - an eight laboratory evaluation of the provisional algal assay procedure bottle test. Department of Environmental Sciences and Engineering, School of Public Health, University of North Carolina, at Chapel Hill, October.

White, L. A., and R. A. MacLeod. 1971. Factors affecting phosphate uptake by *Aerobacter aerogenes* in a system relating cell numbers to ^{32}P uptake. Applied Microbiology 21:520-526.

White, A., P. Handler, and E. L. Smith. 1968. Principles of biochemistry, Chapter 33. McGraw-Hill, New York.

RATES OF CARBON, OXYGEN, NITROGEN, AND PHOSPHORUS CYCLING THROUGH MICROBIAL POPULATIONS IN STRATIFIED LAKES

C. F. Cordeiro, W. F. Echelberger, Jr., and F. H. Verhoff[1]

Introduction

In aquatic environments and lakes in particular, there are a large number of biological and chemical species co-existing. There occur various forms of interrelationships between these species. It is the intention of this paper to attempt, through the usage of a mathematical model of the system, to investigate the rates of cycling of a few important elements, viz. carbon, oxygen, nitrogen, and phosphorus through the various components of the system. The model is kept simple by grouping the microbial populations into categories based upon the type of nutrient transformations occurring. Also necessary is a realistic description of the kinetics and stoichiometry of both the microbial and chemical interactions occurring. Interaction with the environment and between subsystems is also to be included.

Using a more realistic description of microbial cell growth, the lake was modeled as consisting of two layers, an upper and a lower. This was then used to simulate seasonal variations of populations and nutrient concentrations.

Since the original work of Volterra and Lotka (as cited by Chapman, 1931) on predator-prey interaction, various investigations, both theoretical and practical, have been made into ecological systems. The theoretical investigators (e.g., Garfinkel, 1967; Smith, 1969; and Verhoff et al., 1972) were usually interested in the mathematical properties of the ecological system like the stability, effects of density dependent factors and number of trophic levels. Practical studies were usually motivated by ecological phenomena in systems (e.g., Bloom et al., 1968; and Chen, 1970) or by the specific ecology of a location (e.g., Parker, 1968) and usually attempted to simulate and predict actual concentrations of nutrients and biological organisms as a function of time and possibly position.

While a number of investigators (Chen, 1970; Parker, 1968; DiToro et al., 1970; O'Brien and Wroblewski, 1972; and Patten, 1968) have developed models considering nutrient movement through microbial population, none has completed the nutrient cycles, e.g., they do not consider exchange of CO_2, N_2 or O_2 with the atmosphere, or do not consider nitrogen fixing organisms. Parker's model follows the passage of the nutrient through the trophic levels up to the fish, and O'Brien and Wroblewski, in an analysis similar to that of Verhoff and Smith (1971), follow a single conserved nutrient through the lower trophic levels. The O'Brien analysis is similar to that of Verhoff except that the hydrology of the continental shelf is included.

DiToro et al. (1970) consider the effects of phosphate and nitrate nutrients upon the dissolved oxygen of natural waters as caused by the growth of algae; the cycles of these nutrients are not completed. Patten (1968) considers the effect of these nutrients upon the growth of algae.

It is believed by the authors that the cycling of nutrients through the microbial populations plays a significant role in dynamics of water quality. Hence to determine the effects of the nutrient cycling, the complete cycle of the nutrients must be mimicked by the model. The first attempt at the construction of such a model, including the cycling of carbon, phosphorus, and oxygen, was reported recently (Verhoff et al., 1971). Chen (1970) partially completed the cycling of nitrogen, phosphorus, and oxygen in a stream model.

In order to quantitatively relate the growth rate of microbial organisms with nutrient concentrations, all the above investigators have used a Monod type expression, some using a multiplication of Monod factors to account for two limiting nutrients (e.g., DiToro et al., 1970). The Monod expression does not represent the dynamics of microbial cell growth and the multiplication of factors does not give the growth rate for multiple limiting nutrients. This paper then includes a synopsis of work done by the authors on microbial cell growth (Verhoff et al., 1972).

[1]C. F. Cordeiro is with the Department of Chemical Engineering, W. F. Echelberger, Jr., is with the Department of Civil Engineering, and F. H. Verhoff is with the Department of Chemical Engineering, University of Notre Dame, Notre Dame, Indiana.

Mechanism of Microbial Cell Growth

It is assumed that every cell is either growing (however slowly) or it is lysed. The mode of growth assumed in this model supposes that the cell growth occurs in a two-step manner; the first step involves the accumulation and possible loss of the limiting substrate by the organism, and the second step involves the ingestion of the limiting substrate by the organism, and possibly a subsequent cell division.

For a simplified version of the cell growth process, consider a viable cell in a growing culture just after a cell division. This cell selectively assimilates or adsorbs the nutrient from the culture medium. The cell reaction systems then convert the nutrients which have been assimilated into protoplasmic material characteristic of the particular organism. This process of assimilation followed by conversion may occur once or several times between cell divisions; the theory to be developed will be the same for all cases with the change of one parameter. After some chemical rearrangement of the protoplasmic material and the production of an increased amount of nuclear substance, the cell ultimately divides producing two new viable cells which will repeat the process.

This process can be put into symbolic form to represent the stoichiometry (conservation of mass) as shown below.

$$A + \eta S_s \underset{r_2}{\overset{r_1 \quad r_3}{\rightleftarrows}} B \rightarrow (1 + \varepsilon) A + \rho_2 (\eta - \varepsilon) S_s \quad . \quad (1)$$

One mass unit of a viable newly divided cell A combines with η mass units of limiting substrate S_s to form $1 + \varepsilon$ mass units of B. One mass unit of B can then release the assimilated nutrients back to A and S_s or can ingest the substrate, with possible cell division, to yield $(1 + \varepsilon)$ mass units of A. If cell division always accompanies ingestion as in some bacteria, ε is approximately equal to one. ρ_2 $(\eta - \varepsilon)$ represents the weight of substrate that may be returned to the medium during ingestion and cell division. As ρ_2 is normally considered near zero without changing the mechanism, it could be assumed that part or most of the ingestion occurs during what has been described as the assimilation step.

From the symbolic form (Equation 1), it can be seen that the rate of growth will depend upon the three rate processes r_1, r_2, and r_3 indicated. For the development of the rate functions, consider A, S_s, and B to be mass concentration, i.e., mass per unit volume of solution. For the rate of absorption of nutrient, two facts should hold.

$$r_1 \; \alpha \; A$$

and

$$r_1 \; \alpha \; S_s - S_c$$

in which

A	=	number of cells
S_s	=	substrate concentration in solution
S_c	=	substrate concentration in the cell
r_1	=	rate of adsorption per unit volume
$\therefore \; r_1$	=	$k_1 A (S_s - S_c) = k_1 AS$

in which

$S = S_s - S_c$ which is almost equal to S_s as S_c could be very small.

The rate loss of limiting substrate per unit volume by ingesting cells B (i.e., rate r_2), and the rate at which B ingests substrate or ingests substrate and divides (i.e., rate r_2 rate) are assumed proportional to B with all other quantities held constant, i.e.,

$$r_2 = k_2 B, \quad r_3 = k_3 B$$

k_1, k_2, and k_3 are assumed to be independent of all other changing substances in the solution.

Spatial Divisions in Lake

The lake is modeled as consisting of two well-mixed bodies of water, an upper layer and a bottom layer. Each of these will be considered as a separate subsystem with interconnections between them. They are well-mixed, and for the present study, flow inputs are taken to be zero. The upper waters are considered aerobic in nature and are exposed to sunlight giving rise to the possibility of photosynthetic growth of algae together with aerobic bacteria. In the bottom waters only bacterial action is postulated to be occurring. The species in each of these systems can be classified into two categories, biological and chemical (organic or inorganic). In the upper waters (Figure 1) the inorganic chemical species can be further classified into volatile or non-volatile. The volatile ones are those which are capable of being transferred into the system from the atmosphere or out of it into the atmosphere. In this model, these consist of oxygen, nitrogen, and bicarbonate. The non-volatile components are NH_4^+, NO_3^-, and $H_2PO_4^-$. The organic components which have been designated as CPN, CP, CN, and CC are described as follows. CPN has the same composition as the microbial populations. CP is formed on decomposition of CPN with the evolution of NH_4^+ and CN is produced when $H_2PO_4^-$ is removed from CPN. The compound CC is formed when both NH_4^+ and $H_2PO_4^-$ are removed from CPN. The biological species in the upper waters consist of three different groups of algae and three of bacteria. These are grouped according to their function and their nutrients and do not represent particular species as a whole. The first algal group, A_1, uses NH_4^+ as a nutrient; the second, A_2, uses N_2; and the third, A_3, takes up NO_3^-. Each has its corresponding ingestion stage A_{21}, A_{22}, and A_{23}. They also use $H_2PO_4^-$ and HCO_3^- as nutrients. The first form of bacteria, H_1, oxidizes NH_4^+ to NO_3^-. The second, H_2, utilizes both NO_3^- and O_2 to decompose the carbonaceous materials while the last, H_3, uses solely NO_3^-

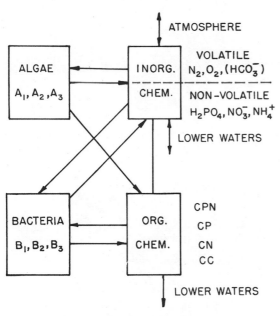

Figure 1. Schematic of upper waters.   Figure 2. Schematic of lower waters.

as an oxygen source reducing it to N_2 in the process. They too have their corresponding secondary stages, H_{21}, H_{22}, H_{23}.

In the bottom waters (Figure 2) the inorganic chemical species are the same with the exception of N_2 and the addition of CH_4. The organic chemicals are the same but for the addition of organic acids which as a group are postulated to have the same composition as CC. They are formed due to the anaerobic decomposition of the organic compounds. There are three forms of bacteria which are modeled as being present in the bottom waters. The first, B_1, produces organic acids which are taken up by the second, B_2, to produce methane gas. B_3, an aerobe is present to scavenge whatever oxygen may be present. Their growth stages are given the nomenclature, B_{21}, B_{22}, and B_{23}.

Only the chemical components are transferred into and out of each subsystem. The organic components move downward through the system via sedimentation; that is, they move from the upper waters through the lower waters into the sediments where they accumulate. The inorganic components are left free to transfer in any direction across the interface between the top and bottom layers according to the concentration gradient.

Kinetics

There are essentially five types of transfer of nutrients in a lake system, 1) convective, i.e., hydrodynamic, 2) diffusive, i.e., simple diffusion of substances, 3) sedimentation, 4) chemical reaction and 5) biological reactions, i.e., those involved in bacteria and algae. The kinetics of the first three types of transfer are understood to some extent and reasonable transfer rate functions can be formulated. In this model, convective transfer occurs in two instances. In the first case, there is an interchange between the atmosphere and the upper waters. It is assumed that transfer of the volatile inorganic components takes place between a thin film on the surface in saturated equilibrium with the atmosphere and the main body of water. The rate of transfer in the case of O_2 for example is given by

$$r_{O2a} = k_{O2a} (Y_{eq.O_2} - Y_{O_2 Upper})$$

(in all the rate equations,

r's	=	mass accumulations or masses reacted of the components per unit time per unit volume of the system
k's	=	mass transfer coefficients or reaction constants

113

Y's = mass concentrations of the components in the system)

Mass transfer is also the mechanism used for transfer between the upper and lower layers. The transfer rate from a unit volume of the upper waters is then given by

$$r_{O2b} = k_{O2b} (Y_{O2 \ upper} - Y_{O2 \ lower})$$

in the case of O_2 and is similarly given in the case of the other inorganic chemical species. The rate of transfer into a unit volume of the lower waters is the same except for a factor corresponding to the ratio of the volumes of the two layers.

Sedimentation, or settling, is assumed to account for the net downward transfer of the organic chemical species. The rate of settling out of a unit volume of the upper waters is given by a fraction of the concentration in that volume. Thus, the rate of settling of CPN from a unit volume of the upper waters is

$$r_{SCPN} = k_{SCPN} Y_{CPN \ upper}$$

The rate of settling into a unit volume of the lower waters would again involve multiplication by the ratio of the two layer volumes. A similar expression is used for sedimentation from the lower waters to the bottom.

Chemical reactions account for the decomposition of the organic components into the simpler species. These reactions are assumed to be first order, e.g., the reaction involving the decomposition of CPN to CN has a rate denoted by

$$r_{P_1} = k_{P_1} Y_{CPN}$$

The algal and bacterial kinetics for both the upper and lower waters are somewhat more complicated since a dynamic description of microbial cell growth as discussed previously was used. This mechanism essentially divides the algal and bacterial masses into two parts, the absorbing part and the ingesting part. The rate functions describing the absorption of nutrients from the waters are described by formulae written in terms of mass changes in the ingesting parts.

For example, for the algae group A_1 the rate would be

$$r_{A1a} = k_{A11} Y_{A1} Y_{H2PO4} Y_{HCO3} Y_{NH4}$$

and for the bacteria H_2 in the upper waters.

$$r_{H2a} = k_{H12} Y_{H2} Y_{O2} (Y_{CNP} + N_1 Y_{CP} Y_{NO3} + N_2 Y_{CN}$$

$$+ N_3 Y_{CC} Y_{NO3})$$

The bacterial rate formulation has four terms to indicate the usage of the four organic substrates.

For the bacterial group B_1 in the bottom waters the rate is given by the formula

$$r_{B1} = k_{B11} Y_{B1} (Y_{CPN} + N_1 Y_{CP} Y_{NH4} + N_2 Y_{CN}$$

$$+ N_3 Y_{CC} Y_{NH4})$$

and for the group B_2 the rate is given by

$$r_{B2} = k_{B12} Y_{B2} Y_{OA} Y_{NH4} Y_{H2PO4}$$

The disappearance by ingestion of the algae and bacterial components are assumed to be first-order in their mass concentrations. For example

$$r_{A21i} = k_{A21i} Y_{A21}$$

is the ingestion rate for component A_{21}. The lysing rates also follow the same first order law.

Stoichiometry

Stoichiometric relationships are important in order to conserve the nutrients. They are required for the absorption, ingestion, and lysing processes of the algae and bacteria, and are also set up for the decomposition of the organic materials.

In order to set up these relationships, it is necessary to know the elemental composition of the microorganisms and the organic compounds present in the system. Various investigators have come up with differing compositions. The chemical formula for mixed algal cultures in surface waters and mixed bacterial cultures in waste treatment processes has been proposed to be $C_{106} H_{180} O_{45} N_{16} P$ by Stumm and Tenney (1963). Redfield (1958), and Eckenfelder (1956), McCarty (1965), Speece and McCarty (1964), Hoover and Porges (1952), and Burkhead and McKinney (1969) conclude that a typical empirical formula for bacterial cells is $C_5 H_7 O_2 N$. Extrapolation to the same N to P ratio in the previous formula yields the formula $C_{80} H_{112} O_{32} N_{16} P$, while keeping the same C to P ratio (105 to 1) gives $C_{105} H_{145} O_{42} N_{21} P$ for the composition of the algae and bacterial cells. A study of the effect of the composition in aerobic systems has been done by Verhoff et al. (1971). If the organic component CPN is assumed to have the composition $C_a H_b O_c N_d P$, then the decomposition into CP would be given by the formula.

$$(C_a H_b O_c N_d P)_{CPN} \rightarrow (C_a H_{b-4d} O_c P)_{CP} + dNH_4^+$$

with similar formulations for the other decomposition reactions.

114

The algal reaction involves the adsorption of the inorganic species and as an example the stoichiometry for the reaction for group A1 is

$$(C_a H_b O_c N_d P)_{A1} + \eta_1 (aHCO_3^- + dNH_4^+ + H_2PO_4^-$$

$$+ ((b-a-2-4d)/2)H_2O) \rightarrow A_{21}$$

Bacteria on the other hand takes up the organic species and may stoichiometrically require some inorganic species. For instance, for the upper water group H_2 the reaction is

$$(C_a H_b O_c N_d P)_{H2} + \begin{bmatrix} CNP \\ CP + d(NO_3^- + 2H_2O - \frac{5}{2}O_2) \\ CN + H_2PO_4^- \\ CC + d(NO_3^- + 2H_2O \\ \quad - \frac{5}{2}O_2 + H_2PO_4^-) \end{bmatrix} \rightarrow$$

$$(\epsilon_2 + (\eta_2 - \epsilon_2)S_{H2})$$

$$+ (\eta_2 - \epsilon_2)(1 - S_{H2})(3a + 4 + 3d$$

$$+ (b - a - 2)/2 - c)/2)O_2 \rightarrow H_{22}$$

For the bottom waters bacteria B_1 the stoichiometry of adsorption is

$$(C_a H_b O_c N_d P)_{B1} + \eta_{B1} \begin{Bmatrix} CPN \\ CP + dNH_4^+ \\ CN + H_2PO_4^- \\ CC + dNH_4^+ + H_2PO_4^- \end{Bmatrix} \rightarrow$$

$$(1 + \eta_{B1})B_{21}$$

and to the bacteria B_2 it is

$$(C_a H_b O_c N_d P)_{B2} + \eta_{B2}(1 - S_1 - S_2)\left[C_a H_{b-4d-2}O_{c-4}\right]_{OA}$$

$$+ (\eta_{B2} - \epsilon_{B2})S_H \left[\frac{1}{3}(4a - b - c + 4d + 6)H_2O\right]$$

$$+ \epsilon_{B2}(S_1 + S_2)\left[dNH_4^+ + H_2PO_4^-\right] \rightarrow$$

$$(1 + \eta_{B2} + (S_H - S_1 - S_7)(\eta_{B2} - \epsilon_{B2}))B_{22}$$

These stoichiometric relations are due essentially to mass balances showing the mass of nutrients which combine with a unit mass of the organism to form the ingesting stage.

The ingestion stoichiometry for the algae and bacteria in the above examples is then given by the following expressions:

$$A_{21} \rightarrow (1 + \epsilon_1)(C_a H_b O_c N_d P)_{A1}$$

$$+ (1 - S_{A1})\eta_1 (3a + 4 + (b - a - 2 - 4d)/2 - c)O_2$$

$$B_{22} \rightarrow (1 + \epsilon_2)(C_a H_b O_c N_d P)_{B2}$$

$$+ (\eta_2 - \epsilon_2)\left[aHCO_3^- + dNO_3^- + H_2PO_4^-\right.$$

$$\left. + (b - a - 2)/2H_2O)\right]$$

For the bottom waters bacteria,

$$B_{21} \rightarrow (1 + \epsilon_{B1})B_1 + (\eta_{B1} - \epsilon_{B1})(1 - S_1 - S_2)$$

$$\left[C_a H_{b-4d-2}O_{c-4}\right]_{OA}$$

$$+ (\eta_{B1} - \epsilon_{B1})(S_1 + S_2)\left[dNH_4^+ + H_2PO_4^-\right]$$

and

$$B_{22} \rightarrow (1 + \epsilon_{B2})B_2 + (\eta_{B2} - \epsilon_{B2})S_{HC}$$

$$\left[\frac{1}{9}(4a - b + 2c + 4d - 6)HCO_3^-\right]$$

$$+ (\eta_{B2} - \epsilon_{B2})S_{CH}\left[\frac{1}{9}(5a + b - 2c - 4d + 6)CH_4\right]$$

In the above equations several stoichiometric constants are used which are defined as follows:

$$S_{A1} = \frac{\text{mol. wt. of organism } A_1}{\text{wt. of corresponding nutrients}} = \frac{AM}{ANH}$$

$$S_{H2} = \frac{\text{mol. wt. of organism } B_2}{\text{wt. of corresponding nutrients}} = \frac{AM}{AN}$$

$$S_1 = \frac{d(\text{ionic wt. of } NH_4^+)}{\text{mol. wt. of CPN}} = \frac{18d}{AM}$$

$$S_2 = \frac{\text{ionic wt. of } H_2PO_4^-}{\text{mol. wt. of CPN}} = \frac{97}{AM}$$

$$S_H = \frac{18(4a - b - c + 4d + 6)}{3(AM)}$$

$$S_{HC} = \frac{61(4a - b + 2c + 4d - 6)}{(AM)}$$

$$S_{CH} = \frac{16(5a + b - 2c - 4d + 6)}{9(AM)}$$

in which

$$AM = 12a + b + 16c + 16d + 3$$
$$AN = 61a + 62d + 97 + 9 \, (b\text{-}a\text{-}2)$$
$$ANH = 61a + 18d + 97 + 9 \, (b\text{-}a\text{-}2\text{-}4d)$$

Light and temperature variations play an important part in the growth rate of both the algae and the bacteria. The growth rate of algae was assumed to be proportional to the light intensity.

$$\text{e.g.,} \quad k_{A21} = k^o_{A21} \cdot \frac{L}{L_s}$$

in which L is the actual intensity of light and L_s is the maximum intensity.

$$\frac{L}{L_s} = \frac{L_o}{L_s} \exp \, [(-\alpha - \beta \Sigma P)H]$$

in which

$$
\begin{aligned}
L_o &= \quad \text{incident light intensity} \\
\alpha &= \quad \text{background/ft} \\
\beta &= \quad \text{algal suspension/ft-mg/} \ell \\
H &= \quad \text{effective depth of algal suspension}
\end{aligned}
$$

$$\frac{L_o}{L_s} = P_p \cdot \frac{L_D}{L_s}$$

in which

$$P_p = \quad \text{photoperiod in days/day}$$

and

$$
\begin{aligned}
L_D &= \quad \text{daily incident light on surface} \\
P_p &= \quad P_{pMax} \, (A_p + (1\text{-}A_p) \, \text{Sin} \, (\phi \text{-} \phi_p)) \\
P_{pMax} &= \quad \text{maximum photoperiod} \\
\phi_2 &= \quad 2 \, t/360; \, t = \text{day of the year} \\
\phi_p &= \quad \text{lag of maximum photoperiod} \\
P_{pMax} \, (2A_p &\text{-}1) = \text{minimum photoperiod} \\
L_D/L_s &= \quad \frac{\text{average daily intensity}}{\text{maximum light intensity}} \\
&= \quad (A_L + (1\text{-}A_L) \, \text{Sin} \, (\phi \text{-} \phi_L))
\end{aligned}
$$

in which

$$\phi_L = \quad \text{lag of maximum light intensity}$$
$$(2A_L\text{-}1) = \text{minimum ratio of intensities}$$

The temperature profile was similarly modeled resulting in an equation of the form.

$$T = T_{max} (A_T - (1 - A_T) \, \text{Sin} \, (\phi - \phi_T))$$

in which

$$\phi_T = \quad \text{lag of maximum temperature}$$
$$T_{max} \, (2A_T\text{-}1) = \text{minimum temperature}$$

The bacterial reproduction rate that is assumed to be linearly varying with temperature has the form

$$k_{B2} = k^o_{B2} \, T/T_{max}$$

The transport rates and the saturation concentrations were also made temperature dependent.

Simulation of the Model

The mass balances performed on a unit volume of the upper waters resulted in 22 differential equations, and those on the bottom waters yielded 17 differential equations to be solved simultaneously. These were run to a steady state using constants which were chosen realistically. Aeration rates, maximum biological growth rates, and sedimentation rates were within the range of values measured and used by others. Other coefficients were estimated from available data. Lake parameters were chosen to correspond to those of Stone Lake near Cassopolis, Michigan.

Results

It is the main intention of this paper to study the rates of transport of the major nutrients between the various components of the system. It is desired to see the major sources of these nutrients and the components which use most of each element. This would give an idea as to which species may be affected most by changes in the system and enable a rough qualitative prediction of the results of corrective actions on the system. The four major elements involved in the growth of the microorganisms are carbon, oxygen, nitrogen, and phosphorus. From the modeled dynamics of the system, it is possible to study the transport of the major compounds of these elements through the system.

Looking at the transfer rates for oxygen (Table 1), it is evident that a major source is the algae which released it at a high rate in the summer. Bacteria can be seen to consume a large portion of the oxygen produced by the algae. The only time when oxygen is transferred from the air to support the bacteria is in the late fall. There is a net scrubbing effect of oxygen in the bottom waters by the third group of bacteria B_3 which is high in both summer and winter.

Taking the HCO_3^- rates into consideration (Table 2), bacteria supply a major portion of that needed by the algae. In the summer (August), especially, the bacteria supply more than twice the carbon than does the atmosphere. That the rate of production of CO_2 cannot be used as a useful guide to bacterial action may be seen from the fact that though the major growth periods are in the late summer and early fall, CO_2 is released to the atmosphere only in the late winter through early summer periods. This is due to a combination of algal uptake and saturation effects. The major transfer from top to bottom occurs in the cooler months and upwards in the warmer months.

$H_2PO_4^-$ (Table 3) is used as the major phosphorus ion in the system for simplicity and also because it is the ion which occurs in the largest quantities. Phosphorus is also one of the elements which is essentially not replenished from the atmosphere and must rely on runoff or

Table 1. Transfer rates of O_2.

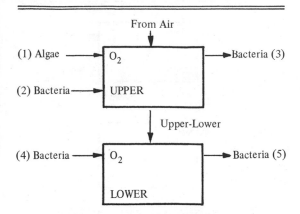

Transfer Rates (mg per liter upper waters per day)

Transfer Component	February	May	August	November
From Air	-0.003	-0.44	-0.27	0.26
Upper-Lower	0.006	0.028	0.005	0.01
1	0.14	0.62	0.99	0.48
2	0.014	0.02	0.02	0.04
3	-0.09	-0.14	-0.80	-0.79
4	0.015	0.006	0.006	0.018
5	-0.019	-0.025	-0.014	-0.033

Table 2. Transfer rates of HCO_3^-

Transfer Rates (mg per liter upper waters per day)

Transfer Component	February	May	August	November
From Air	-0.186	-0.169	0.40	0.193
Upper-Lower	0.022	-0.48	-0.034	0.277
1	0.345	0.187	0.96	1.02
2	-0.12	-0.92	-1.37	-0.6
3	-0.027	-0.066	-0.033	0.045

other inputs. It would thus seem to lend itself to easier controllability and its function in the lake system is therefore of interest. Bacterial action in the model supplies a major portion of the phosphorus needed for the high summer algal growth. In the late winter and early summer, however, the chemical decomposition of the organic species yields phosphorus at a greater rate than that of the bacteria. The net transport is upwards in the summer when there is high algal uptake and downwards in the winter. In the bottom waters, the $H_2PO_4^-$ produced by bacterial decomposition almost equals that taken up for stoichiometric balance.

Nitrogen is also an important element in the composition of the microorganisms. In this model, three forms of inorganic nitrogen have been postulated, NH_4^+, NO_3^-, and dissolved N_2.

Ammonia (Table 4) production in the upper and bottom waters is mainly by the chemical decomposition of the organic material. It can be seen that a higher uptake of NH_4^+ by algae occurs in the early summer (May) than in the late summer. The concentration of this algae reaches a peak earlier than the other forms. Very little is taken up by the bacteria in the upper waters as contrasted to the bottom waters. Transport of NH_4^+ is mainly upwards due to excess production in the bottom waters. NO_3^- (Table 5) production on the other hand is by the oxidation action of the bacteria. It is more than sufficient to supply the needs of the algae which are highest in the late summer. The consumption of NO_3^- in the bottom waters by bacteria is higher than its production because the NO_3^- ion acts as an election acceptor. The net transfer of NO_3^- is towards the bottom as the major production occurs in the upper layers.

The algal concentration profiles (Figure 3) show that the algal group utilizing NH_4^+ is much more productive than that using NO_3^- as is observed in many cases. The rate of accumulation of bottom sediment (Figure 4) was calculated using the formula.

$$D = 12x \cdot \frac{V}{A} \cdot \frac{1}{\rho} \text{ inches}$$

Table 3. Transfer rates of $H_2PO_4^-$

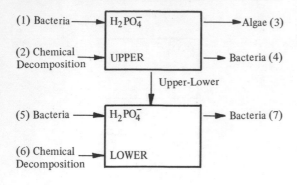

Transfer Rates (mg per liter upper waters per day)

Transfer Component	February	May	August	November
Upper-Lower	0.0023	-0.004	-0.0054	0.0017
1	0.006	0.003	0.015	0.016
2	0.009	0.006	0.004	0.006
3	-0.002	-0.015	-0.02	-0.009
4	-0.0075	-0.0045	-0.0038	-0.0056
5	0.0010	0.0029	0.0015	0.0011
6	0.003	0.0011	0.0008	0.0008
7	-0.0013	-0.0036	-0.0013	-0.0011

Table 4. Transfer rates of NH_4^+.

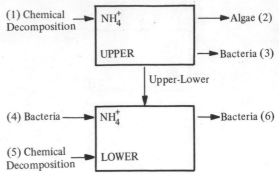

Transfer Rates (mg per liter upper waters per day)

Transfer Component	February	May	August	November
Upper-Lower	0.0037	-0.0021	-0.0114	0.0047
1	0.022	0.013	0.007	0.011
2	-0.002	-0.026	-0.018	-0.010
3	0.00	0.0	0.0	0.0
4	0.002	0.007	0.003	0.001
5	0.011	0.019	0.006	0.005
6	-0.004	-0.014	-0.006	-0.001

Figure 3. Algal concentration profiles.

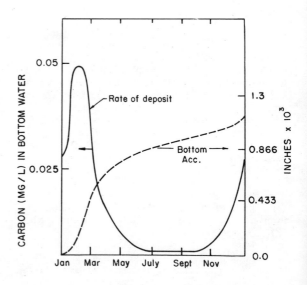

Figure 4. Rate of bottom deposit and net accumulation.

118

Table 5. Transfer rates of NO_3^-.

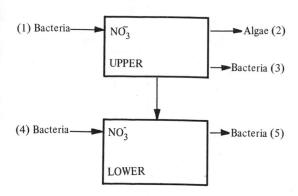

Transfer Rates (mg per liter upper water per day)

Transfer Component	February	May	August	November
Upper-Lower	0.0007	0.0011	0.0040	0.0037
1	0.218	0.0069	0.145	0.160
2	-0.006	-0.019	-0.109	-0.035
3	-0.211	-0.049	-0.025	-0.124
4	0.0034	0.0045	0.0024	-0.0059
5	-0.0039	-0.0057	-0.0046	-0.0124

in which

x = rate of sedimentation of organics from the bottom waters in mg/ℓ-day
V = volume of bottom waters in acre-ft
A = area of bottom in acres
ρ = density of sediment in mg/ℓ

The modeled rate of deposition is highest in the early part of the year and then tapers off. The rate of accumulation is directly proportional to the sedimentation rate. Utilization of a sedimentation rate close to that occurring naturally would give an idea of the time it would take the body of water to silt up.

Summary

The rates of cycling of nutrients are an important part of the dynamics of an aquatic ecosystem. Most rates cannot be measured directly from analysis of water quality parameters. They cannot be easily estimated from concentration values of the chemical and biological species present in the system. Modeling contributes significantly to the understanding of this nutrient cycling. It is through the use of a proper model that these rates can be reasonably estimated. In the construction of such a model, care should be taken that each element cycling through the system is conserved. The kinetics of both the chemical and biological reactions should be realistic, and finally, the stoichiometric relationships of these reactions should be balanced.

The four major elements included in this study are carbon, nitrogen, phosphorus, and oxygen, and the model has been based on the study of these elements and their pertinent compounds. In order to preserve as much simplicity as possible, only a few important forms have been used. The kinetics of cell growth are developed from a mechanism which postulates that this growth is accomplished by two consecutive processes, the transport of metabolite from the exterior of the cell to the interior, and the chemical conversion of these chemicals into cellular photoplasm. Only the rate of the transport process was assumed dependent on the external concentrations of the nutrients, while the conversion rate is held dependent on the light and temperature.

The exact elemental composition of the microbial populations is not known and so an average composition is used based on investigations by other workers. This was used as a basis for deriving the various stoichiometric relationships used in the model.

The lake was simulated as consisting of two layers, each with its set of biological and chemical components. The transport rates between the different components of the major elements indicated some interesting results. For instance, the oxygen (Table 1) requirements for bacterial growth are fully met by algal production except for the winter (November) when it supplies 60 percent and most of the rest (34 percent) is transferred in from the atmosphere. Bacteria supply most of the bicarbonate (Table 2) necessary for algal growth. In the period of highest algal growth (August) 70 percent of the bicarbonate is produced by the bacteria compared to 30 percent from the atmosphere. The temperature effect in saturation concentrations may be observed in May when, even though algae take up bicarbonate at a rate higher than bacterial production, there is a release into the atmosphere, and from the lower to the upper waters due to the decrease of the equilibrium saturation concentrations with the increase in temperature.

Bacteria also supply 75 percent of the $H_2PO_4^-$ used by the algae in the late summer (August) while in the early summer (May) the phosphate requirements are met more by the chemical decomposition (40 percent) as compared to the bacteria (20 percent). The rate of algal uptake of NH_4^+ is higher in May (0.026 mg/ℓ-day) than in August (0.018 mg/ℓ-day) as contrasted to NO_3^- which has a rate of 0.019 mg/ℓ-day in May and 0.109 mg/ℓ-day in August.

The symbiotic activity of the algae and bacteria as regards to oxygen and bicarbonate production and utilization is clearly evident in the model. It is also possible to study the sources and sinks of the other elements. The settling and silting rate can also be determined from the model. Such a model readily lends itself to prediction of responses to control actions.

References

Bloom, S. G., A. A. Levin, and G. E. Rains. 1968. Mathematical simulation of ecosystems—a preliminary model applied to a lotic freshwater environment. Battelle Memorial Institute, Columbus, Ohio.

Burkhead, C. E., and R. E. McKinney. 1969. Energy concepts of aerobic microbial metabolism. Proc. Amer. Soc. Civil Engrs. 95(SA2):253.

Chapman, R. N. 1931. In Animal Ecology. McGraw-Hill, N. Y., p. 409.

Chen, C. W. 1970. Concepts and utilities of ecologic model. Proc. Amer. Soc. of Civil Engr. 96(SA5):1083.

Cooney, C. L., and D. I. C. Wang. 1970. Influence of environmental conditions of microbial cell growth: Experimental and mathematical analysis. Presented at 63rd Annual Meeting AIChE, Chicago.

DiToro, D. M., J. J. O'Connor, and R. V. Thomann. 1970. A dynamic model of phytoplankton in natural waters. Enviro. Eng. Sci. Program., Manhattan College, Bronx, N. Y.

Eckenfelder, W. W., and R. F. Weston. 1956. Kinetics of biological oxidation. In Biol. Treatment of Sewage and Industrial Wastes. Vol. 1, Rheinhold, New York, p. 18.

Garfinkel, D. 1967. A simulation study of the effect upon simple ecological systems of making the rate of increase of population density dependent. J. of Theor. Biol. 14:46.

Hoover, S. R., and N. Porges. 1952. Pilot plant for treatment of Kraft mill wastes. Sewage and Ind. Wastes 24:306.

McCarty, P. L. 1965. Thermodynamics of biological synthesis and growth. In 2nd Int. Conf. on Water Pollution Research. Vol. 2, Pergamon Press, New York, p. 169.

Monod, J. 1949. The growth of bacterial culture. Ann. Reo. Microbiol. 3:371.

O'Brien, J. J., and J. S. Wroblewski. 1972. An ecological model of the lower marine trophic levels on the continental shelf off West Florida Coast. Masters Thesis, Florida State University.

Parker, R. A. 1968. Simulation of an aquatic ecosystem. Biometrics 24:803.

Patten, B. C. 1968. Mathematical models of plankton production. Int. Rev. ges. Hydrobiol. 53:570.

Redfield, A. C. 1958. The biological control of chemical factors in the environment. Amer. Sci. 46:205.

Smith, F. E. 1969. Effects of enrichment in mathematical models. In Eutrophication: Causes, Consequences, Correctives. Nat. Acad. Sci., Washington, D.C., p. 631.

Speece, R. E., and P. L. McCarty. 1964. Nutrient requirements and biological solids accumulation in anaerobic digestion. In First Int. Conf. on Water Pollution Research. Vol. 2, Pergamon Press, London, p. 305.

Stumm, W., and M. W. Tenney. 1963. Waste treatment for control of heterotrophic and autotrophic activity in receiving waters. Twelfth Southern Municipal and Industrial Waste Conference, Raleigh, N. C.

Tenney, M. W., W. F. Echelberger, P.C. Singer, F. H. Verhoff, and W. A. Garvey. Fourth Int. Conf. on Water Pollution Research, Pergamon Press (to be published).

Verhoff, F. H., W. F. Echelberger, Jr., M. W. Tenney, P. C. Singer, and C. F. Cordeiro. 1971. Lake water quality prediction through system modeling. Proc. 1971, Summer Computer Simulation Conference, Boston, Mass., p. 1014-1023.

Verhoff, F. H., and F. E. Smith. 1971. Theoretical analysis of a conserved nutrient ecosystem. J. Theor. Biol. 33:131-147.

Verhoff, F. H., K. R. Sundaresan, and M. W. Tenney. 1972. A mechanism of microbial cell growth. Biotech. and Bioeng. 14:411.

CAPABILITIES AND LIMITATIONS OF A NUTRIENT-PLANKTON MODEL[1]

R. A. Parker[2]

Introduction

During the last decade a large number of plankton production models have been developed. These have ranged in complexity from those embedded in massive ecosystem models (Jansson, 1972) to others of moderate scope (DiToro et al., 1970; Parker, 1968; Steele, 1971), as well as many of a more limited nature (see review by Patten, 1968). Every model represents a compromise between the reality of a natural system and the simplicity of a mathematical description. Prediction is the usual goal, utilizing sufficient biological and physical-chemical detail in construction to enhance reliability over a broad environmental spectrum. Fitted "regression" equations often compare favorably in predictive capacity with causally based functions; however, they provide little insight into the underlying processes involved. Regardless of the approach, a detailed evaluation of the course of eutrophication suggests a strategy of model formulation which focuses on nutrient-plankton relationships.

Of particular interest in understanding the consequences of nutrient enrichment is, of course, the algal growth response. Early models (Patten, 1968) frequently assumed that the rate of photosynthesis depended on the concentration of a single nutrient in a linear fashion. More recently a Monod or Michaelis-Menten dependence has been adopted for single or multiple nutrients (e.g. DiToro et al., 1970; Parker, 1973). "Half-saturation" constants vary considerably among species and nutrients. Values have been measured by several investigators under a variety of experimental conditions (Eppley et al., 1969; MacIsaac and Dugdale, 1969; Thomas and Dodson, 1968). Since ammonium may be used preferentially as a nitrogen source by certain phytoplankters, the purpose of the study reported here was to evaluate the performance of a coupled system of nutrient and plankton differential equations which incorporated this assumption in contrast to one without preference.

The Model

The basic model has been described previously (Parker, 1973). It was developed in an effort to determine the effects of inorganic phosphate effluent entering Kootenay Lake in southeastern British Columbia. This large body of water (area over 400 square kilometers and 150 meters deep) received substantial quantities (to 15 metric tons daily) between 1953 and 1969 from a fertilizer plant located on the Kootenay River. In the original formulation, photosynthetic activity of two assemblages of algae was made to depend on phosphate concentration (Nl) as the function $Nl/(K_1 + Nl)$, together with nitrate (N2) and ammonium (N3) in functional form $(N2 + N3)/(K_2 + N2 + N3)$. Note that the two nitrogen sources were really treated as one. K_1 and K_2 were set equal to 1 and 6 micromoles per liter, respectively. Although the multiplication of Monod expressions to describe the effect of multiple nutrients has been questioned (particularly by Verhoff, personal communication), this dependence was retained because of widespread usage in other studies. Table 1 summarizes eight variants chosen for analysis, Trial I being that reported in Parker (1973).

Table 2 reviews the growth and mortality functions adopted for the algal groups, and for the two zooplankton groups (cladocerans and copepods). Vollenweider (1970) has concentrated on one comprehensive relationship between photosynthesis and light intensity; however, a somewhat simpler expression averaged over depth was utilized here having an optimum intensity of 1.6 calories per square centimeter per hour for both groups in the range 400-700 nannometers. Temperature optima of 20° and 11°C were chosen based on field observations with $\alpha_1 = 4.80$ and $\alpha_2 = 1.45$. Values for other constants can be found in Parker (1973).

The entire system of differential equations employed is presented in Table 3. Horizontal transport is recognized for each variable, and vertical diffusivity has been included for the three nutrients. These hydrodynamic components are extremely important, difficult to describe (Huber and Harleman, 1968; Sweers, 1970; Wilson and Marsh, 1967), and the entire model is very sensitive to their contributions. Only the upper "mixed" layer was modeled, limited to 10 meters so that virtually all photosynthetic activity was included. Constant

[1]This project has been financed in part with Federal funds from the United States Environmental Protection Agency under grant number R-800430.

[2]R.A. Parker is with the Departments of Zoology and Computer Science, Washington State University, Pullman, Washington.

121

nutrient concentrations in the massive hypolimnion were maintained with changes allowed only at the time of fall overturn (the lake ordinarily circulates continuously throughout the winter).

Photosynthetic rate, nutrient uptake, and cell nutrient content may or may not all exhibit Monod dependence on external nutrient concentration. Here the nutrient content was taken to be a constant proportion of the total biomass produced. Returning to Trial I in Table 1, it is seen that the phosphate content of both algal groups was set to 0.3 micromole per milligram and the total nitrogen content to 2.0 micromoles per milligram (a relatively low N/P ratio). Growth constants a_{11} and a_{21} in Table 2 were 35 and 8, respectively. Trials II and III considered the effect of the independent action of nitrate and ammonium with the same half-saturation constants for both algal groups; Trial II treated growth dependence on these two nutrients equally, whereas Trial III made growth much more sensitive to low ammonium concentrations. Trial IV allowed algal group 1 to grow as in Trial III, and algal group 2 as in Trial I. Cell content was arbitrarily (and perhaps controversially) taken to be *inversely* proportional to the half-saturation constant.

Growth constant values were adjusted to account for differences in the magnitudes of the various growth functions. For example, F21•F31 for group 1 in Table 1 has a mean value of about one-half of that for F2+3 over concentrations from 0 to 6; therefore, the growth constant was raised to 70 from 35. Talling (personal communication) has questioned the relatively high half-saturation constant of 1.0 for phosphate, hence Trials V through VIII were conducted using the model as in Trials I-IV but with a phosphate half-saturation constant of 0.1.

Results and Discussion

No attempt was made to maximize fit to field data by parameter estimation (as in Parker, 1972); rather, attention is called to the seasonal pattern of results from the numerical solution of the differential equation system. In the original discussion (Parker, 1973), simulation output was presented for each of four horizontally spaced sampling stations. Here the only output given will be for station 2 located nearly 11 kilometers from station 1 at the river mouth. System activity at this station (2) is the most difficult to simulate since no lake plankton is assumed to be contributed by the river, whereas, input

Table 1. Nutrient-growth functions,[a] algal contents (µm/mg), growth constants, and error sums of squares.

Trial	Algal Group	Growth Function N1	N2	N3	Content N1	N2	N3	Growth Constant	Error SS
I	1	F11	F2+3		0.30	2.00		35	511
	2	F11	F2+3		0.30	2.00		8	
II	1	F11	F21	F31	0.30	1.00	1.00	70	403
	2	F11	F21	F31	0.30	1.00	1.00	16	
III	1	F11	F22	F32	0.30	0.33	1.67	70	698
	2	F11	F22	F32	0.30	0.33	1.67	16	
IV	1	F11	F22	F32	0.30	0.33	1.67	70	699
	2	F11	F2+3		0.30	2.00		8	
V	1	F12	F2+3		0.30	2.00		35	795
	2	F12	F2+3		0.30	2.00		8	
VI	1	F12	F21	F31	0.30	1.00	1.00	70	491
	2	F12	F21	F31	0.30	1.00	1.00	16	
VII	1	F12	F22	F32	0.30	0.33	1.67	70	934
	2	F12	F22	F32	0.30	0.33	1.67	16	
VIII	1	F12	F22	F32	0.30	0.33	1.67	70	962
	2	F12	F2+3		0.30	2.00		8	

[a]$F11 = N1/(1. + N1)$ $F21 = N2/(3. + N2)$ $F31 = N3/(3. + N3)$ $F2+3 = (N2 + N3)/(6. + N2 + N3)$
$F12 = Nl/(.1 + N1)$ $F22 = N2/(5. + N2)$ $F32 = N3/(1. + N3)$

from one lake station contributes to the next and aids mathematically in timing plankton pulses.

Simulation results for phytoplankton, phosphate, nitrate, ammonium, copepods, and cladocerans are presented for a three-year period (1966-69) in Figures 1-6, respectively. Although optimization in terms of goodness-of-fit was not attempted, the squared relative error summed over the six variables for all sampling dates is given in Table 1. Only total chlorophyll was measured;

however, simulated values for both algal groups are shown in Figure 1. Note the typical spring pulse (group 2) followed by a peak in late summer and early fall (group 1). There is a general depression of phosphate during the summer (Figure 2), nitrate depletions corresponding to phytoplankton pulses (Figure 3), and relatively low ammonium levels throughout the period from late spring until late fall (Figure 4). As one would expect, phosphate never reaches the point where it limits production (in contrast to nitrate). High ammonium concentrations

Table 2. Growth and mortality rates used in model.

| | Growth Rate (G) | Mortality Rates | |
		Natural (M_1)	Predation (M_2)
Algal Group 1 (A_1)	$a_{11}\bar{e}_1(I)f_1(T)g_1(N_1)h_1(N_2,N_3)$	$a_{21}T$	$TB(c_{11}C_1 + c_{21}C_2)$
Algal Group 2 (A_2)	$a_{21}\bar{e}_2(I)f_2(T)g_2(N_1)h_2(N_2,N_3)$	$a_{22}T$	$TB(c_{12}C_1 + c_{22}C_2)$
Cladocera (C_1)	$b_1TB(c_{11}A_1 + c_{12}A_2)$	$c_{13}T$	$c_{14}TC_1$
Copepoda (C_2)	$b_2TB(c_{21}A_1 + c_{22}A_2)$	$c_{23}T$	$c_{24}TC_2$

$$I = \text{light intensity}$$
$$T = \text{temperature}$$
$$N_i = \text{nutrients}$$
$$e_i(I) = (I/I_{si})\exp(1 - I/I_{si})$$
$$f_i(T) = [(T/T_{si})\exp(1 - T/T_{si})]^{\alpha_i}$$
$$g_i(N_1) = N_1/(K_{1i} + N_1)$$
$$h_i(N_2,N_3) = (N_2 + N_3)/(K_{2i} + N_2 + N_3)$$
$$B = \exp[-k(A_1 + A_2)^r]$$

Table 3. Differential equation model.

$$A_1' = A_1(G_{A_1} - M_{1A_1} - M_{2A_1}) - v\,\partial A_1/\partial x$$

$$A_2' = A_2(G_{A_2} - M_{1A_2} - M_{2A_2}) - v\,\partial A_1/\partial x$$

$$C_1' = C_1(G_{C_1} - M_{1C_1} - M_{2C_1}) - v\,\partial C_1/\partial x$$

$$C_2' = C_2(G_{C_1} - M_{1C_2} - M_{2C_2}) - v\,\partial C_2/\partial x$$

$$N_1' = p(A_1M_{2A_1} + A_2M_{2A_2} - C_1G_{C_1} - C_2G_{C_2}) - p(A_1G_{A_1} + A_2G_{A_2}) + m\,\partial N_1/\partial z - v\,\partial N_1/\partial x$$

$$N_2' = -n_1(A_1G_{A_1} + A_2G_{A_2})N_2/(N_2 + N_3) + m\,\partial N_2/\partial z - v\,\partial N_2/\partial x$$

$$N_3' = n_1(A_1M_{2A_1} + A_2M_{2A_2}) - n_2(C_1G_{C_1} + C_2G_{C_2})\ n_1(A_1G_{A_1} + A_2G_{A_2})N_3/(N_2 + N_3) + m\,\partial N_3/\partial z - v\,\partial N_3/\partial x$$

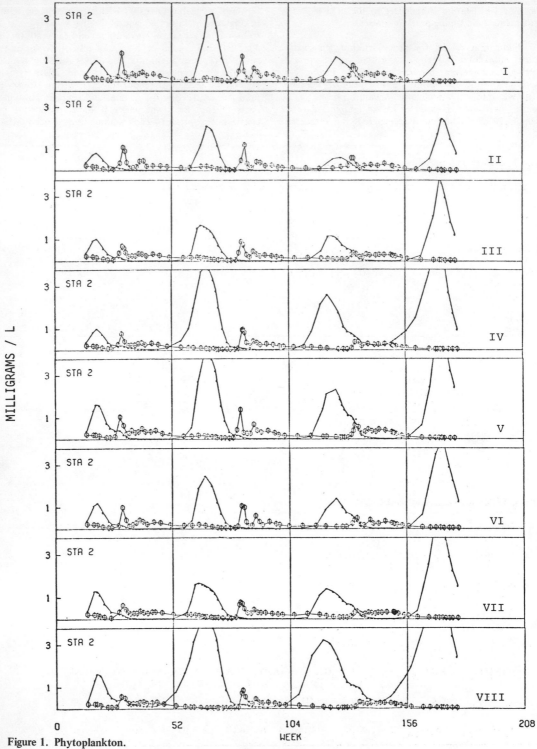

Figure 1. Phytoplankton.

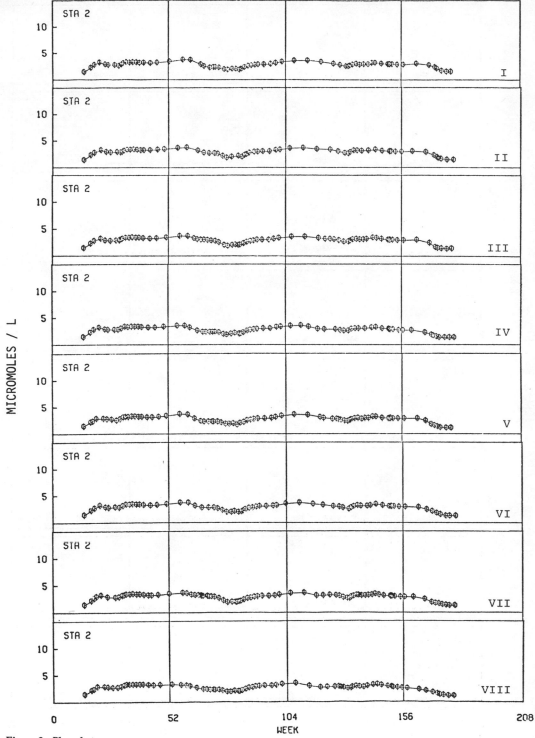

Figure 2. Phosphate.

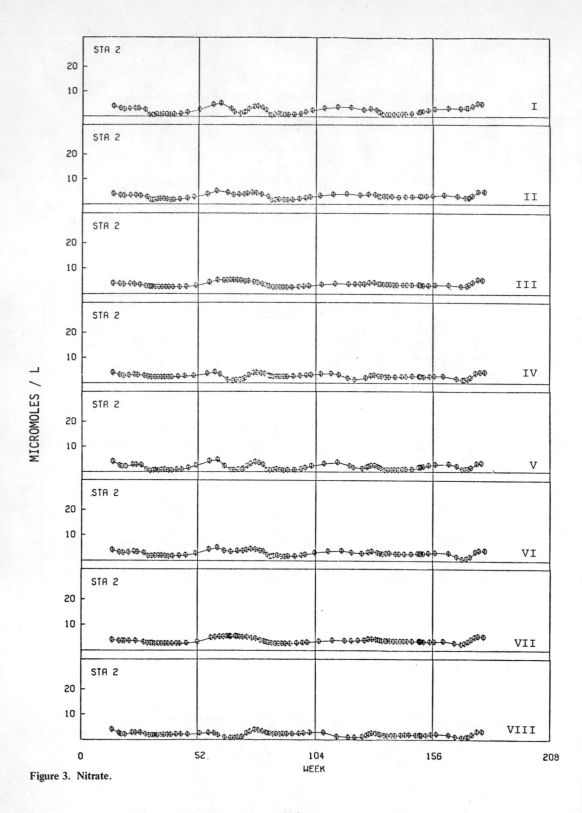

Figure 3. Nitrate.

126

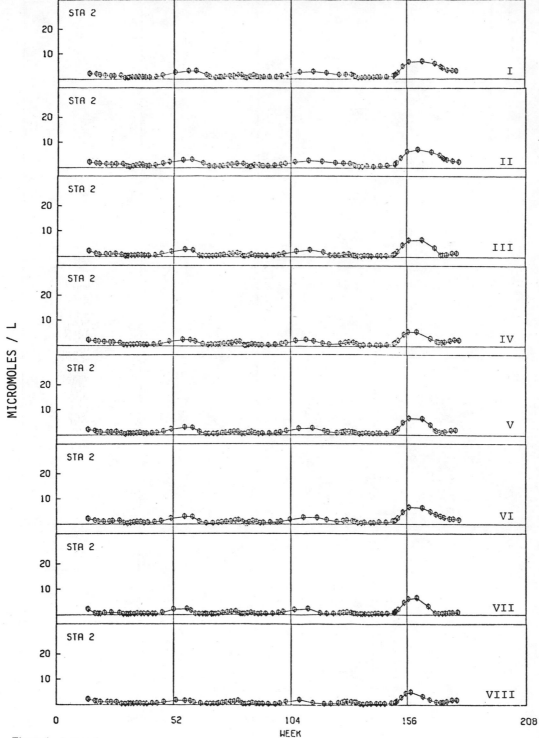

Figure 4. Ammonium.

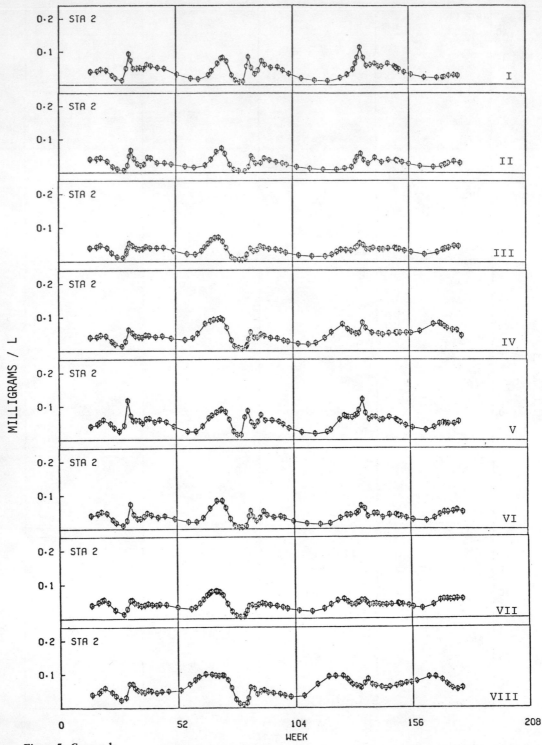

Figure 5. Copepods.

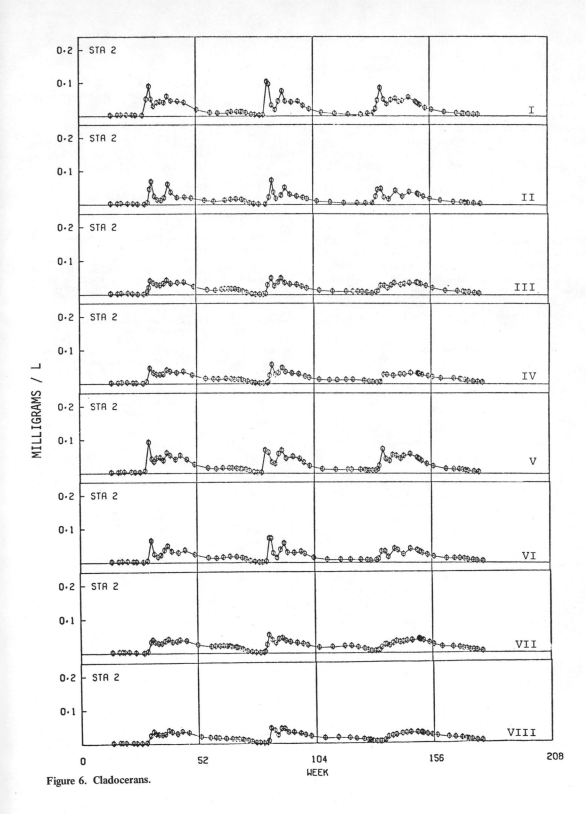

Figure 6. Cladocerans.

129

during the 1968-69 winter were caused by an abnormally large influx via the Kootenay River. Copepods are present during the entire year with peak population densities following phytoplankton pulses (Figure 5). Cladocerans are abundant in August and September (Figure 6), somewhat higher in September than one usually finds in the Lake. This difference may be due to extensive predation by land-locked salmon, a species not included in the model due to lack of observational data.

Aside from the notable reduction in the error sum of squares for Trial II, it seems reasonably apparent that the model does not depend heavily on modifications in the Monod nitrate-ammonium growth functions, nor on the half-saturation constant for phosphate. The fundamental seasonal patterns are the same for all eight versions. One is led to conclude that the performance of *this model* is under the over-riding influence of a structure inherently dependent on temperature and light intensity. Continued evaluation based upon additional knowledge of growth kinetics may shed more light on the roles of various nutrients in regulating phytoplankton productivity in aquatic systems subject to accelerated eutrophication.

Literature Cited

DiToro, D.M., D.J. O'Connor, and R.V. Thomann. 1970. A dynamic model of phytoplankton populations in natural waters. Environmental Engineering and Science Program, Manhattan College.

Eppley, R.W., J.N. Rogers, and J.J. McCarthy. 1969. Half-saturation constants for uptake of nitrate and ammonium by marine phytoplankton. Limnol. Oceanogr. 14(6):912-920.

Huber, W.C., and D.R.F. Harleman. 1968. Laboratory and analytical studies of the thermal stratification of reservoirs. Rept. No. 122, Hydrodynamics Laboratory, MIT, Cambridge. 277 p.

Jansson, B-O. 1972. Ecosystem approach to the Baltic problem. Bull. 16, Ecol. Res. Committee, Swedish Nat. Sci. Res. Council. 82 p.

MacIsaac, J.J., and R.C. Dugdale. 1969. The kinetics of nitrate and ammonia uptake by natural populations of marine phytoplankton. Deep-Sea Res. 16(1):45-57.

Parker, R.A. 1968. Simulation of an aquatic ecosystem. Biometrics 24(4):803-821.

Parker, R.A. 1972. Estimation of aquatic ecosystem parameters. Verh. Internat. Verein. Limnol. 18:257-263.

Parker, R.A. 1973. Some problems associated with computer simulation of an ecological system. *In* Mathematical Theory of the Dynamics of Biological Populations (M.S. Bartlett and R.W. Hiorns, eds.). Academic Press, London.

Patten, B.C. 1968. Mathematical models of plankton production. Int. Rev. Ges. Hydrobiol. 53(3):357-408.

Steele, John H. 1971. Factors controlling marine ecosystems. *In* Nobel Symposium 20—The Changing Chemistry of the Oceans. (D. Dyrssen and D. Jagner, eds.), pp. 209-221. John Wiley and Sons, New York.

Sweers, H.E. 1970. Vertical diffusivity coefficient in a thermocline. Limnol. Oceanogr. 15:273-281.

Thomas, W.H., and A.N. Dodson. 1968. Effects of phosphate concentration on cell division rates and yield of a tropical oceanic diatom. Biol. Bull. 134(1):199-208.

Vollenweider, R.A. 1970. Models for calculating integral photosynthesis and some implications regarding structural properties of the community metabolism of aquatic systems. *In* Prediction and Measurement of Photosynthetic Productivity. pp. 455-472. Proc. IBP/PP Technical meetings, Trebon 14-21 September 1969.

Wilson, J.R., and F.D. Marsh. 1967. Field investigation of mixing and dispersion in a deep reservoir. Tech. Rept. HYD 10-6701, Hydraulic Engineering Laboratory, University of Texas, Austin. 141 p.

PHYTOPLANKTON POPULATION CHANGES AND NUTRIENT

FLUCTUATIONS IN A SIMPLE AQUATIC ECOSYSTEM MODEL

R. R. Lassiter and D. K. Kearns[1]

Introduction

Aquatic ecosystems, inherently difficult for terrestrial man to investigate, are the subject of investigation by diverse and ingenious techniques. One technique being used is the formulation of complex views of the system into working mathematical models. These models serve both as statements of a hypothesis about mechanisms by which a system functions and as initial tests of that hypothesis.

The mathematical model here was purposefully oversimplifed because its initial purpose was to lead into the construction of a more comprehensive model. While serving its initial purpose, it is providing insights, perhaps largely because it is oversimplified. It may be thought of as representing a large flask culture or at best a small pond during a period when there is no inflow.

The model has provided an opportunity to experiment with ecological processes which involve interactions between chemical and biological systems. In fact any experiment which influences one system necessarily influences the other. Experimentation with this simple system has provided results which have influenced plans for modeling more complex systems.

Overview of the Model

The model consists of equations describing a physico-chemical system and an associated biota. Physical aspects of the system consist of light, temperature, and an atmosphere containing 0.033 percent CO_2. Chemical aspects consist of a set of dissolved inorganic and organic chemicals. The inorganic chemicals are in chemical equilibrium. It is assumed that the organic chemicals do not participate directly in the chemical equilibria. The biotic

system consists of six phytoplankton species and heterotrophic microorganisms. The model contains explicit equations for the phytoplankton, but the heterotrophic microorganisms are included only implicitly by the assumption that non-living organic material decomposes at a rate proportional to its concentration. Algal growth rates are temperature dependent.

The physical and chemical systems interact by CO_2 exchange at the air-water interface, thereby providing input into the system of dissolved chemicals. Also temperature affects the rate of gaseous exchange and the CO_2 saturation value.

Assumptions of this model do not provide for true interaction between the physical and biotic systems. The physical system effects a change in the phytoplankton growth rate, as noted, and light is required for phytoplankton growth; but there is no effect of biotic system upon physical system. One noteworthy omission of such an effect is the reduction of available light by the phytoplankton.

The chemical and biotic systems interact in several ways. Direct interactions are the uptake of inorganic chemicals by phytoplankton, the production of organic materials by the biota via excretion and death, and conversion of organic chemicals to inorganic chemicals by microorganisms.

Equations of the Model

Light

It is assumed that algal growth is proportional to light intensity. This is an oversimplification, but is approximately correct for light levels not greatly exceeding the optimal intensity. An expression for growth as a function of light may be substituted in place of the linear dependence used here.

Because of the linear dependence and an additional assumption of independent action of all growth-modifying factors, a value in the interval (0,1) was computed and

[1] R. R. Lassiter and D. K. Kearns are with the National Pollutants Fate Research Program, Southeast Environmental Research Laboratory, National Environmental Research Center-Corvallis, U.S. Environmental Protection Agency, Athens, Georgia. Mr. Kearns' current address is University of Georgia, Athens, Georgia.

used as one of the factors. This value in the interval (0,1) was computed using the expression

$$Y = Max \left\{ 0, \left[sin \left(\frac{\pi}{12} t \right) + .25 \right] 1.25 \right\}. \quad (1)$$

Here t is time in hours. There are 24 hours per 2π radians, hence the argument of the sine. The value .25 is added to increase the day length to about 14 hours, and the divisor, 1.25, scales Y to the (0,1) interval. Figure 1 illustrates this function graphically.

Temperature

Temperature is described in the model by a sine wave scaled to a period of 365 days. The range of temperature is 7°C to 35°C.

Three types of temperature expressions are used in the model. The first describes the effect of temperature upon the exchange rate of CO_2 between atmosphere and water. This expression is derived from the empirical expression for the oxygen reaeration rate, usually written as

$$k_T = k_{25} \theta^{(T-25)} \quad \ldots \ldots \ldots \ldots (2)$$

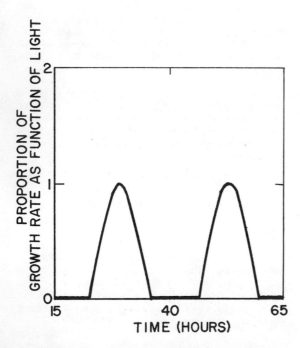

Figure 1. Growth rate as function of light intensity for two daylight cycles.

in which
    T is temperature, degrees Celsius,
    θ is a constant, and
    k is a reaeration rate at the temperature indicated
        by its subscript.
The value of θ is reported variously as 1.024 (Churchill et al., 1962) and 1.022 (Tsivoglou, 1973). Equation 2 may be written also as

$$k_T = k_{25} e^{c(T-25)} \quad \ldots \ldots \ldots \ldots (3)$$

the form used in the computer program for this model. Evaluation of the constant, c, gives .022 (Tsivoglou's) and .024 (Churchill's).

Temperature influences the saturation value of CO_2 in water. Using the data of Bohr (Hutchinson, 1957, p. 654) the expression

$$C_{S,T} = A e^{bT}$$

was fit, with the resulting specific equation

$$C_{S,T} = 2.28 \times 10^{-5} e^{-.318T} \quad \ldots \ldots (4)$$

in which $C_{S,T}$ is the saturation value of CO_2 in water at temperature T (°C), and it is expressed as moles/liter.

The influence of temperature upon biological growth rates is described by

$$k_T = k_{opt} e^{a(T-T_{opt})} \left\{ \frac{T_{max}-T}{T_{max}-T_{opt}} \right\}^{a(T_{max}-T_{opt})} \quad \ldots (5)$$

in which
    k_T is the biological rate,
    T_{opt} is optimal temperature,
    K_{opt} is the rate when temperature is optimal,
    T_{max} is the temperature above which growth rate,
        $k_T = 0$, and a is a constant which embodies
        the effect of a unit deviation of T from T_{opt},
        a unit deviation of T from T_{max}, and the rate
        constant for the change in k with T.

The rationale, derivation, comparison with other functions, etc., of Equation 5 will be presented in a paper in preparation. It is sufficient here to say that the function provides a good representation of a wide variety of growth versus temperature data, and that graphically, it has the general appearance indicated in Figure 2.

The chemical system

The chemical system used is comprised of two subsystems: Inorganic and organic. The organic subsystem consists of three components: Organic carbon, organic nitrogen, and organic phosphorus. These are artificial categories, separated for convenience in studying the transient fate of the nutrients and to allow atomic ratios

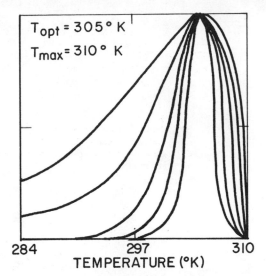

$T_{opt} = 305° K$

$T_{max} = 310° K$

FAMILY OF CURVES FOR
a = (.1, .2, .5, 1, 2) (OUTER TO
INNER)

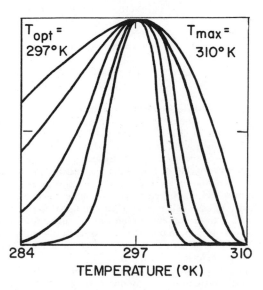

$T_{opt} = 297°K$

$T_{max} = 310°K$

FAMILY OF CURVES WITH
SAME VALUES OF a AS ABOVE

BIOLOGICAL RATE AS FUNCTION OF TEMPERATURE

Figure 2. Biological rate as function of temperature for two temperature optima and five proportionality constants each.

of assimilated phytoplankton biomass to be maintained. In this model the algae excrete organic carbon, which therefore varies with phytoplankton growth activity. In more complex growth models, all organic fractions may vary much more than in this model.

The organic subsystem is linked to the inorganic by decomposers transforming organic material to inorganic. In all simulations with this model it has been assumed that carbon, nitrogen, and phosphorus are transformed from organic to inorganic at the same rate. The computer program allows for different decomposition rates to be used if desirable.

Equations for the organic components are of the form:

$$\frac{dO_j}{dt} = A + S - D \quad \dots \dots \dots \dots (6)$$

in which

O_j = concentration of the jth organic component

A = input rate from phytoplankton deaths

= $\sum_{i=1}^{n} f_{ji} m_i P_i$

in which

n = number of phytoplankton populations

f_{ji} = mole fraction composition of the jth chemical element in the ith phytoplankton population

m_i = death rate of the ith phytoplankton population

P_i = biomass of the ith phytoplankton population

S = secretion input rate

= $\sum s_{ji} G_i$

in which

s_{ji} = secretion rate of the jth chemical element by the ith phytoplankton population

G_i = growth rate of the ith phytoplankton population [see Equation 8]

D = decomposition rate loss of organic to inorganic material

= $o_j O_j$

in which

o_j = decomposition rate

The inorganic subsystem is comprised of two types of equations: Differential equations for losses and gains of inorganic materials, and algebraic equations defining the (assumed) instantaneously achieved chemical equilibria. Gains are (for carbon) through the air-water interface, and (for all) via decomposition as noted. Losses are (for carbon) through the air-water interface, and (for all) by phytoplankton uptake. The form of the differential equation is

$$\frac{dI_j}{dt} = E + D - U \quad \cdots \cdots \cdots \cdots \quad (7)$$

in which

I_j = concentration of the jth inorganic component

E = term for gaseous exchange at the air-water interface

= $k_j (I_{js} - I_j)$

in which

k_j = exchange rate of the jth inorganic component, temperature dependent, see Equation 3

I_{js} = saturation concentration of the jth inorganic species, temperature dependent, see Equation 4

D = decomposition as defined for Equation 6

U = loss rate of uptake by phytoplankton population i of the jth inorganic chemical

$$= \sum_{i=1}^{n} (f_{ji} + s_{ji})\, G_i$$

in which the symbols are as defined for Equation 6 under terms A and S

The differential equations for gains and losses in combination with algebraic equations for equilibrium provide the system with the capability of maintaining proper distribution of the chemical forms.

The algebraic equations are of two types: Distribution fractions and a net charge equation. The distribution fractions are computed from the equilibrium equations of the form

$$\frac{[H^+][Y^-]}{[HY]} = K$$

and are functions of $[H^+]$ for all equilibria considered in this model. More complex equilibria may be computed where the distribution fractions are functions of two or more variables. Other computing schemes are required for the more complex chemical equilibria.

The net charge equation is simply the algebraic sum of positive and negative charges. In computing the equilibria, $[H^+]$ is varied until a set of inorganic chemical species is found which makes the net charge sufficiently close to zero. This set of species is considered the equilibrium set. The system is similar (with a few elaborations) to the system presented on page 183 of Stumm and Morgan (1970).

Phytoplankton

The phytoplankton equations include terms for increase and decrease of biomass. Their general form is

$$\frac{dP_j}{dt} = G - A \quad \cdots \cdots \cdots \cdots \quad (8)$$

in which

P_j = concentration of biomass of phytoplankton species j

G = growth rate of phytoplankton population j

$$= \hat{\mu}_j \left\{ \prod_{i=1}^{C} I_i / (K_{ji} + I_i) \right\} \cdot P_j \cdot L_j$$

in which

$\hat{\mu}_j$ = specific growth rate maximum for phytoplankton species j, a function of temperature [see Equation 5]

I_i = concentration of inorganic chemical i

K_{ji} = half-saturation constant of the jth phytoplankton species for nutrient inorganic chemical i

L_j = effect of light intensity upon species j, (Equation 1) all species are identical with respect to this factor in this model

A = phytoplankton death rate, more completely defined in Equation 6

Several observations are in order regarding Equation 8. First, although several authors (Rich et al., 1971; O'Connor et al., 1973; Chen and Orlob, 1971; Schofield, 1971) have utilized a model which has this form with respect to the nutrients, there seems to be little rationale for its use. That is, it does not seem easily derivable using meaningful statements about algal physiology as a starting point. It has been criticized as resulting in values which are too small when more than one factor is suboptimal, e.g., if three nutrients are present in concentrations equal to the half saturation constant for each, the growth rate would be no greater than .125 $\hat{\mu}$ even though no factor alone repressed growth below .5 $\hat{\mu}$. Second, it may be interpreted to mean that the rate of growth of phytoplankton is a function of the external concentration of each nutrient independent of all others. Ketchum (1939) represented data which showed lack of this dependence for the diatom *Nitschia closterium,* and the fact that the organisms are manufacturing materials requiring all three chemicals argues for dependence. Third, although the equation deals with multiple nutrients, it affords no capability for uptake or growth as a function of nutrients internal to the cells, a concept which has been argued for by Caperon (1967) and Koonce (1972), and which is the only intuitively plausible concept.

However, because it is a simple model which affords the opportunity to work with a system wherein there are multiple nutrients, it has been used in this research on an interim basis until a more satisfactory model is more completely checked.

Discussion of Results Obtained Using the Model

The model presented here, though not large scale by every standard, is complicated somewhat by the presence of both chemical equilibria and dynamic equations. Experiments with the model thus far have been to observe behavior in these two major subsystems.

Model tests

Some of the criteria used to test this model's capability and correctness are charge balance for the chemical equilibria, proper partitioning of the forms of chemicals, conservation of mass where applicable, fluctuations of pH in response to removal of inorganic carbon by the phytoplankton.

Net charge was set at $|10^{-14}|$ moles, a value found to be attainable and to preserve mass over a fairly long simulation interval. Test computer runs showed that after the initial interval, only one to four iterations were required for the modified Newton-Raphson routine to converge to within $|10^{-14}|$ moles.

Checks in various sources showed that the forms of inorganic carbon, ammonia nitrogen, and orthophosphate were properly partitioned with respect to pH.

In this model no provision was made for net gains or losses of ammonia nitrogen or orthophosphate. Thus it was possible to check these two components for conservation of mass. Adding the quantities partitioned into algal biomass, dead organic material, and inorganic material, each of which varied dynamically, a total was carried throughout the simulations. For all runs thus far, total phosphorus has deviated from the initial total by a maximum of seven parts in about 10,000, while total nitrogen has changed less than one part in about 10,000.

Experimentation with the model

As with most mathematical models, profitable experimental use of the model requires a different and perhaps more difficult mode of thought. Because of this, experimentation proceeds more slowly and less generally than design and construction of the model. Experimentation thus far with this model may be characterized as exploratory.

Prior to discussion of specific sets of results, two general observations can be made based upon the exploratory use of the model. The first observation has to do with rates. Because growing seasons are finite, and because rates may be submaximal due to the lack of perfect matching of niche requirements to environmental conditions, the timing of growth may be such that a population of phytoplankton may never reach the potential set by total system nutrients. Rates of secretion and decomposition of organic materials are two system parameters which are important in establishing timing of growth. Second, algal blooms as depicted by the model, are accompanied by a pattern of transient changes in the forms in which the nutrients exist. In order for a bloom to develop rapidly, there must be a large supply of nutrients available in inorganic form. Such a bloom converts a large proportion of the inorganic nutrients to algal biomass. Termination of the bloom (in this model usually by depletion of one or more nutrients below available levels) is followed by increase in the concentration of non-living organic materials. Finally bacterial mineralization of the nutrients occurs. Although all these processes are continuously occurring together in the model, the series (inorganic-algal-organic-inorganic ...) described has recurred in cases of simulated algal blooms. The key to this behavior may be the size of the inorganic nutrient pool at the beginning of the growing season, and as importantly, the types of organisms present to take advantage of the situation. Species succession will be discussed in this sense following a discussion of simpler experiments on development of nutrient limitation.

Development of nutrient limitation.

Several sets of results were obtained wherein a single phenomenon recurred. The phenomenon was the development of phosphorus limitation. This phenomenon is easily explained. The model represents a system closed to all materials except carbon. The atomic ratio of assimilated nutrients in algal biomass was fixed at 100:15:1. The total nitrogen to total phosphorus ratio in the system was about 40:1, and all nutrients were recycled from organic to inorganic form at the same rate. Thus it is clear that phosphorus, required at 1/15 the nitrogen rate (nutrient assimilated per unit increase in algal biomass) but present at only 1/40, will become limiting before nitrogen. Where experiments were begun with low inorganic carbon such that there was carbon limitation, there was always a continued buildup of carbon in the system until finally phosphorus became limiting. As noted earlier, high rates of secretion of organic carbon served to extend the time period wherein phosphorus was not limiting. The diel pH pattern is strikingly correlated with the occurrence of phosphorus limitation. Prior to limitation of growth by low phosphorus concentration, pH fluctuated a varying amount depending upon the set of factors controlling uptake of CO_2. At the time when phosphorus became limiting, pH fluctuations terminated, being subsequently controlled by purely chemical and physical factors (Figure 3). Further experimentation will be required to determine whether there are key features to the pH patterns, which could be used as indicators of what factors are critical in a

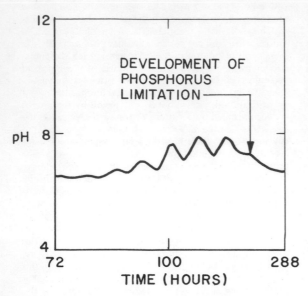

12

DEVELOPMENT OF
PHOSPHORUS
LIMITATION

pH 8

4
72   100   288
TIME (HOURS)

Figure 3. pH variations during a simulated algal bloom. Increasing amplitude indicates greater algal growth rates; termination of the fluctuations accompanies development of phosphorus limitation.

given situation. Rich et al. (1971) have reported, for example, the prediction of characteristic pH fluctuations using a model simulating a situation where carbon is limiting. It appears from preliminary experimentation with this model that pH fluctuates noticeably only when there is significant phytoplankton growth. Fluctuations resulting from carbon limitation are seen in this model because it assumes no uptake of CO_2 at night (hence a pH decrease) and uptake during the day (hence a pH increase) to whatever extent the interaction of available carbon and assimilation rate allow. Because the system is open to carbon but closed to other nutrients, this phenomenon appears transiently.

Species succession

An assumption basic to this research is that development of eutrophic conditions, as determined by presence of large blooms of blue-green algae and the replacement of "desirable" by "undesirable" algal species is a result of species succession. Given this assumption, it is important to determine how changes in chemical and physical factors of a water body influence species succession. It was hypothesized that succession cannot be predicted based upon nutrient load alone, but that it results from

the complex of factors which dynamically alter environmental conditions, thereby continuously changing competitive advantages.

One experiment has been done thus far toward studying species succession. For the discussion the six phytoplankton species will be called S_1, S_2, S_3, S_4, S_5, and S_6. Here the factors temperature, maximal growth rate ($\hat{\mu}$), and nutrient uptake capability with respect to concentration (K, half saturation constant) were set so that growth response to these factors could be observed. S_1, S_3, and S_5 were given high $\hat{\mu}$ values (.35 hr^{-1}), and high K values (5 x 10^{-8} mole liter^{-1}). S_2, S_4, and S_6 were given relatively lower values, $\hat{\mu}$ = .25 hr^{-1}, and K = 1 x 10^{-8} mole liter^{-1}. These growth rate values are extremely high as will be apparent if they are expanded to their day^{-1} value. However, there are several reasons why they were set this high. First, they must be high because of the characteristic of the model to reduce the growth rate more than is realistic when more than one influencing factor is suboptimal. Second, day length is only 14 hours long so that expansion from hr^{-1} to day^{-1} requires multiplication by 14 rather than 24. Finally they were purposefully made high for most experiments thus far to make any effects show up more distinctly. When phenomena are understood qualitatively, then quantification may proceed more easily.

The temperature optima were set so that the organisms appear in pairs, one of each pair with low $\hat{\mu}$ and K values and one with high. They were set also so that the pairs were rather widely separated by temperature optima. The optima were S_1 and S_2 about 18°C, S_3 and S_4 25°C, S_5 and S_6 32°C. Temperature followed the pattern discussed earlier, reaching a maximum of 35° about 2160 simulated hours from time zero.

Irregular patterns of algal population changes are observed in natural systems. Figure 4, solid curve, shows the irregular pattern of simulated algal biomass resulting from this model. The general similarity of this curve to naturally observed phytoplankton changes suggests a potential of models incorporating species succession to predict seasonal patterns of phytoplankton population changes.

The simulated successional pattern

In Figure 4 it can be seen that S_1 made up nearly the entire initial bloom. The peak occurred between days three and four and was terminated when inorganic nutrients became depleted by the growing organisms. Death processes transferring algal biomass to organic material and further decomposition to inorganic nutrients gradually brought inorganic nutrients back to a level where further growth could occur. Concurrently, temperature increased to where S_3 and S_4 were favored. In this case S_3 with the high growth rate, was out-competed by S_4, with capability to utilize nutrients in low concen-

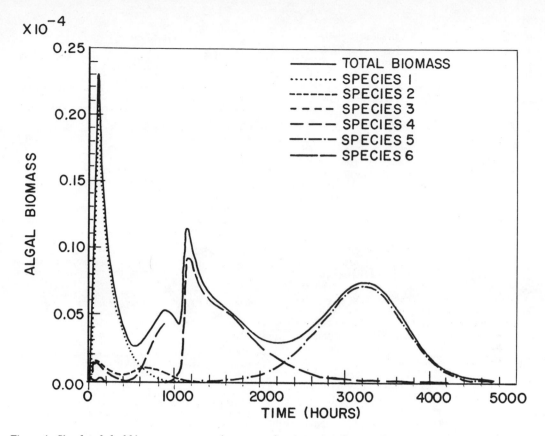

Figure 4. Simulated algal biomass over a growing season showing seasonal succession.

trations. The reason for this is that inorganic nutrients were building up from a very low level, and S_4 was able to start growing while the nutrient level was still too low for S_3. S_4, removing nutrients about as fast as they were made available, never allowed them to reach levels where S_3 could grow. S_4 never reached high population levels because the large concentration of organic matter resulting from the early bloom was decomposed slowly enough that temperature limited the growth. By the time the large organic pool from the S_1 and S_4 blooms were mostly decomposed, the temperature neared the optimum for S_5 and S_6. S_6 reached a rapid population peak by outcompeting S_5 in the same manner as was described above for S_4 over S_3. The bloom by S_6 was also temperature limited when temperature reached levels well above its optimum. Finally after temperature peaked and returned to near optimum levels for S_5 and S_6, inorganic nutrients made available from the high temperature dieoff of S_6, were utilized by S_5 in a late summer bloom that persisted until temperature terminated it well into the simulated "fall."

This narrative of the successional pattern could represent a general description of the diatom-green-bluegreen seasonal pattern often observed naturally. If it

is representative, the model study points up certain features of the system which may be important in nature.

First, the most obvious feature of the simulations is that temperature plays a very large part in determining the pattern. This is not a new observation, but its importance to modeling is that the fidelity of the function relating rates to temperature can determine success or failure of the modeling effort.

Second, whenever nutrients are present largely in available inorganic form (e.g., during early spring), those organisms with high potential growth rates have the competitive advantage. Late in the growing season when biologically cycled nutrients are primarily responsible for growth, a premium is placed upon the uptake capabilities of the organisms.

A third feature of the simulations may not represent the real world. In Figure 4, the last peak represents organisms with a high temperature optimum. These organisms persist and no others replace them even though temperature progresses through the optimum for all the other species. The reason why this happens in the model is that the species have been reduced to such low levels that they are not able to reach competitively high population levels during the time interval of favorable temperature. Resting states probably form in nature, thereby placing a lower limit on the population density for algal species. If this is the case, some organisms may appear in two peaks per growing season, and some, such as S_2 here, may reach population peaks in the fall, whereas they are outcompeted by the fast growers in the spring.

Further studies of this sort, with more fully represented nutrient cycles and perhaps higher trophic levels, may be expected to provide insight into the mechanisms permitting algal blooms to develop and help to understand why some blooms persist. More intelligent prevention and control measures may even be suggested by such model studies.

Summary and Conclusions

This model, although simple in many respects, is a useful tool for investigating many aspects of phytoplankton dynamics. Among the results obtained thus far are those which point to timing of significant ecological events (such as phosphorus reaching critically low concentrations) as important in phytoplankton competition. In some cases pH may prove to be a good indication of the time of occurrence of such events.

It was noted that temperature is of primary importance in determining phytoplankton successional patterns.

Finally it was postulated that fast growing phytoplankton forms may dominate the early growing season, while forms not so fast-growing but with capability to use lower nutrient concentrations may dominate later season growth when their ability to utilize nutrients equals or exceeds the rate of bacterial remineralization.

References

Caperon, J. 1967. Population growth in micro-organisms limited by food supply. Ecology 48(5):715-721.

Chen, C. W., and G. T. Orlob. 1971. Ecologic simulation of aquatic environments. First Annual Report to Office of Water Resources Research, U.S. Department of the Interior.

Churchill, M. A., H. L. Elmore, and R. A. Buckingham. 1962. Prediction of stream reaeration rates. J. San. Engrg. Div., Proc. Am. Soc. Civ. Engrs., 88(SA4).

Hutchinson, G. E. 1957. A treatise on limnology. Vol. 1, Geography, physics, and chemistry. John Wiley and Sons, Inc., New York.

Ketchum, B. H. 1939. The development and restoration of deficiencies in the phosphorus and nitrogen composition of unicellular plants. J. Cell. Comp. Physiol. 13:373-381.

Koonce, J. F. 1972. Seasonal succession of phytoplankton and a model of the dynamics of phytoplankton growth and nutrient uptake. Ph.D. Dissertation, University of Wisconsin, Madison.

O'Connor, D. J., R. V. Thomann, and D. M. DiToro. 1973. Dynamic water quality forecasting and management. U.S. Environmental Protection Agency Final Report for Research Project No. R-800369 (in press).

Rich, L. G., J. F. Andrews, and T. M. Keinath. 1971. Mathematical models as aids in interpreting eutrophication phenomena. Clemson University.

Schofield, W. R., Sr. 1971. A stochastic model of a dynamic eco-system in a one-dimensional, eutrophic estuary. Ph.D. Dissertation, Virginia Polytechnic Institute and State University, Blacksburg.

Stumm, W., and J. J. Morgan. 1970. Aquatic chemistry. Wiley-Interscience, New York.

Tsivoglou, E. C. 1972. Characterization of stream reaeration capability. U.S. Environmental Protection Agency.

AQUATIC MODELING IN THE EASTERN DECIDUOUS FOREST BIOME, U.S.-INTERNATIONAL BIOLOGICAL PROGRAM[1]

J. A. Bloomfield, R. A. Park, D. Scavia, and C. S. Zahorcak[2]

Introduction

The U.S. International Biological Program is organized at the biome level and includes large-scale multi-disciplinary efforts in the Eastern Deciduous Forest, Grassland, Desert, Coniferous Forest and Tundra Biomes. Historically, each group has developed its own modeling capability with slightly different goals and approaches; with the exception of the Grassland Biome, all have supported comprehensive aquatic projects.

The goals of the Eastern Deciduous Forest Biome, broadly stated, are: 1) To understand better the dynamics of ecosystems, and 2) to be able to predict the consequences of man-induced perturbations. In order to achieve the second goal, strong emphasis has been placed on developing mechanistic aquatic models capable of forecasting the effects of nutrient enrichment, thermal pollution, siltation and other by-products of man's activities. At the same time, mindful of the first goal, it was felt that the models should represent to some degree the ecologic insights gained through the large-scale integrated studies of IBP and that the formal logic of the models should serve as a test of the understanding of whole-ecosystem dynamics.

The principal Biome model, known as CLEAN (Comprehensive Lake Ecosystem ANalyzer), is intended to serve these dual objectives by coupling all generalized aquatic process models developed in the Biome in a computer code that facilitates a variety of applications (Park et al., 1972). However, models of three levels of resolution are being developed. There are high-resolution models of certain processes, such as photosynthesis, that may eventually replace simpler models in CLEAN if it can be demonstrated that substantial improvements in eco-

system simulation would result; and there are greatly simplified ecosystem models, including SIMPLE (SIMPlified Lake Ecosystem) that are being developed for specific management, research and teaching purposes.

The Biome has two lake sites, each with a comprehensive research effort. Lake George, New York, is a long, narrow, moderately deep lake with undeveloped to intensively-developed shoreline areas and concomitant oligotrophic to mildly eutrophic water qualities. Investigators include personnel from Rensselaer Polytechnic Institute, State University of New York, at Albany, Siena College, Skidmore College, Marist College, and Union College. Lake Wingra, Wisconsin, is a small shallow lake within the city limits of Madison and is extremely productive. Site investigators are almost all from The University of Wisconsin. Process research is coordinated between the two sites, and data from Lake Wingra is regularly transmitted to the Lake George site for use in further developing and evaluating CLEAN. In addition to participating in the development of CLEAN, the Lake Wingra group has developed a site-specific model of mixed levels of resolution (WNGRA2) in order to facilitate site experimentation, as reported elsewhere in this symposium.

Model Development

At both sites the development of ecosystem models serves as the focal point for integrative research. Incorporation of detailed process information in the model has been made possible by the close working relationships between aquatic specialists and modelers. Feedback from model analysis has, in turn, provided redirection for research. This is particularly important in studying the effects of eutrophication where interaction among ecosystem components often can lead to counterintuitive results.

Establishment of initial research priorities was guided by conceptual models which defined the scope of the overall program and served to identify key ecosystem components and processes. Sampling stations were located in regions of Lake George and Lake Wingra characterized by different combinations of environmental factors, including water depth, nutrient enrichment and exposure to

[1]Research supported by the Eastern Deciduous Forest Biome Project, International Biological Program, funded by the National Science Foundation under Interagency Agreement AG-199, 40-193-69 with the Atomic Energy Commission—Oak Ridge National Laboratory.

[2]J.A. Bloomfield, R.A. Park, D. Scavia, and C.S. Zahorcak are with the Rensselaer Polytechnic Institute, Troy, New York.

wave and current agitation; this permitted the development and subsequent evaluation of "point" models under a variety of conditions in the two lakes.

Q-mode cluster analysis and ordination of data from the first year of intensive field studies helped to confirm that the sampling stations in Lake George were representative of differing degrees of eutrophication. R-mode cluster analysis and ordination of data from both Lake George and Lake Wingra is helping in the definition of functional groups aggregated at levels consistent with the simulation modeling goals.

Formulation of model equations was accomplished at a series of workshops involving a modeling team from Biome headquarters, site modelers and the respective aquatic specialists. These formulations dictated subsequent efforts in estimating the necessary parameters in the field and laboratory. As parameter estimates and additional process information became available the many constructs were evaluated and changed where warranted.

Unfortunately, because of the difficulty of measuring biologic processes, a range of values, rather than precise estimates, is available for most parameters. Therefore, it has been necessary to calibrate, or tune, the models by varying the parameters within the observed ranges until a "best fit" to a particular set of observed data is obtained. At present the applicability is being investigated of both response-surface methodology and non-linear estimation procedures in assessing the sensitivity of the models to each of the parameters and in determining optimal parameter values.

Models have been implemented in PL/1, FORTRAN and DYNAMO. Initially, PL/1 was used at the Lake George site because of the flexibility of logic, extensive library of functions and full character set. However, its use precludes routine program sharing with other sites, so during the development of CLEAN conversion was made to FORTRAN, with careful avoidance of machine-dependent features; WNGRA2 is also in FORTRAN. DYNAMO has been used in developing SIMPLE at the Lake George site. CLEAN is implemented in interactive mode for the UNIVAC 1108 time-sharing system; it is presently being implemented on the IBM 370/165 so that it can be accessed by Environmental Protection Agency personnel in accordance with an interagency agreement. It is written in a command syntax, including in-line editing of parameters, site constants and initial conditions. The Runge-Kutta-Merson integration technique, with variable step-size, is used. Simulation results can be tabulated or plotted (see Appendix A). WNGRA2 utilizes batch mode; it is more fully described elsewhere in this symposium.

Description of CLEAN

Because it embodies much of the "state of the art" of modeling in the Biome, it is appropriate to examine the

structure of CLEAN in some detail. The present model is formulated as 28 coupled ordinary differential equations, representing most of the biotic components of interest in eutrophication studies (Figure 1).

Careful attention has been given to interactions among the components (Figure 2). Allowances were made in programming the model to accommodate additional components as they are required; these will include submodels for water balance, amphipods, yellow perch, and lake trout. Driving variables include incident solar radiation, water temperature, wind or barometric pressure gradient, nutrient concentrations, and allochthonous dissolved and particulate organic matter. The model represents a m^2 column of water.

CLEAN is programmed in modular form so that one can execute a specific submodel, such as for net phytoplankton, or link any meaningful combination of submodels as required. Subprogram functions exist for each process, such as respiration, facilitating program alterations and resulting in a greatly compressed program since most of the biologic processes occur repeatedly with only changes in parameter values.

Model Components

Producers

Producers are divided into two groups—phytoplankton and macrophytes. Both depend on photosynthesis, but whole-organism processes differ greatly. Therefore, these groups have been modeled separately.

Macrophytes (Appendix B1) can be considered as analogous to terrestrial plants because of the predominance of structural material, particularly stems, the importance of roots, and the active storage and transport of food as carbohydrates. Because of the functionality and the availability of data, the macrophyte model is disaggregated:

$$\frac{d B_m}{dt} = \frac{dL}{dt} + \frac{dR}{dt} + \frac{dS}{dt}$$

in which

B_m	=	macrophyte biomass,
L	=	leaf and stem biomass,
R	=	root biomass, and
S	=	carbohydrate pool

Macrophyte photosynthesis

The maximum photosynthetic rate, Pmax, is limited by suboptimal levels of light, temperature, and nutrients (Equation 1.1). Light availability is decreased by absorption by the lake water, phytoplankton and the macrophyte's own leaves (self-shading). Nutrient availability, particularly of carbon, is limited in part by boundary effects between the leaves and surrounding water; therefore, renewal of carbon is expressed as a function of water current velocity. Other nutrients are derived principally from interstitial water, with uptake through the roots; this process will be included in the model as soon as adequate data are available from ongoing studies.

Macrophyte tissue growth

Growth of leaf and stem tissue begins in the spring on the basis of physiologic time (Equation 1.2) and proceeds asymptotically to an optimal leaf area index as

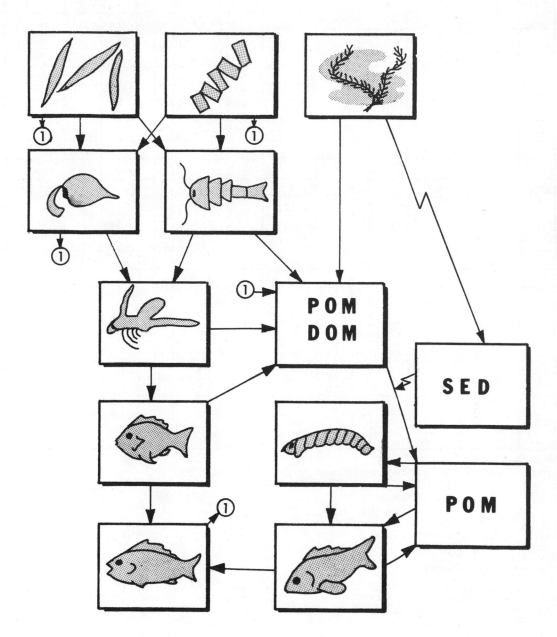

Figure 1. Ecosystem components forming submodels in CLEAN. The Biome aquatic ecosystem model. Decomposers are associated with both particulate organic matter (POM) and dissolved organic matter (DOM).

To	1	2	3	4	5	6	7	8	9	10	11	12	13	14	15	16	17	18	19	20	21	22	23	24	25	26	27	28
1			2.1 4.1	2.1 4.1	2.1 4.1	2.4	2.4										7.2 2.4	7.3 2.3					7.2 2.4	7.3 2.3				
2			2.1 4.1	2.1 4.1	2.1 4.1	2.4	2.4										7.2 2.4	7.3 2.3					7.2 24	7.3 2.3				
3				4.1						4.1 5.1							4.1 7.2	4.4					4.4	4.4				
4				4.1						4.1 5.1							4.1 7.2	4 4					4 4	4.4				
5				4.1						4.1 5.1							4.1 7.2	4.4					4.4	4.4				
6										5.1	5.1	5.1					6.1 7.2	6.1					6.1	6.1				
7											5.1	5.1	5.1				6.1 7.2	6.1					6.1	6.1				
8								6.2	6.2	5.1 6.2	5.1 6.2	5.1 6.2																
9								5.2		5.1 6.2	5.1 6.2	5.1 5.2																
10											5.1						5.3 7.2	5.4					5.3	5.4				
11											5.1						5.3 7.2	5.4					5.3	5.4				
12											5.1						5.3 7.2						5.3	5.4				
13																	7.2											
14																	7.2											
15											1.3	1.5					7.2											
16						6.1	6.1						1.5				7.1	7.1	7.1	7.1	7.1		7.1					7.1
17						7.2 6.1	7.2 6.1					7.2 6.1						7.2					7.2					
18																7.1								7.3				
19	2.2	2.2																										
20	2.2	2.2																										
21	2.2	2.2														7.1												
22						6.1	6.1				5.1											7.1	7.1	7.1		7.1		
23						6.1 7.2	6.1 7.2				5.1						7.2							7.2				7.1
24											5.1	6.1 7.2						7.3				7.1						
25											5.1																	
26											5.1																	
27											5.1																	
28																												

Figure 2. Transfer matrix for CLEAN; numbers refer to equations in Appendix B.

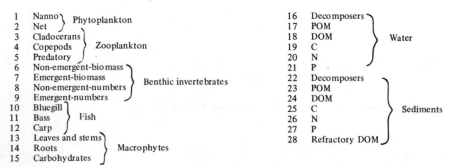

Key to Figure 2

1	Nanno	} Phytoplankton	
2	Net		
3	Cladocerans		
4	Copepods	} Zooplankton	
5	Predatory		
6	Non-emergent-biomass		
7	Emergent-biomass		
8	Non-emergent-numbers	} Benthic invertebrates	
9	Emergent-numbers		
10	Bluegill		
11	Bass	} Fish	
12	Carp		
13	Leaves and stems		
14	Roots	} Macrophytes	
15	Carbohydrates		
16	Decomposers		
17	POM		
18	DOM	} Water	
19	C		
20	N		
21	P		
22	Decomposers		
23	POM		
24	DOM		
25	C	} Sediments	
26	N		
27	P		
28	Refractory DOM		

142

long as available carbohydrate supply from photosynthesis exceeds a minimum level (Equation 1.3). The optimal leaf area index (Equation 1.4) is the surface area of leaves which results in maximum net photosynthesis (O'Neill et al., 1972). Greater leaf surface results in smaller net gain due to self-shading and metabolic cost of maintaining the leaf biomass.

Growth of root tissue proceeds at a rate proportional to the root biomass up to a maximum value as long as there are sufficient available carbohydrates (Equation 1.8).

Macrophyte respiration

Leaf and stem respiration (Equation 1.5) is proportional to leaf and stem biomass and continues as long as active growth is occurring. Root respiration (Equation 1.9) is proportional to root biomass but ceases when available carbohydrates are reduced to a minimal level.

Macrophyte mortality

Mortality occurs in the sense that leaves and, at times, stems are sloughed off (Equation 1.6). Provision is made for sloughing at two periods during the year, although the mechanisms triggering this process are not well understood. Sloughing also occurs when the optimum leaf area is exceeded.

The phytoplankton model (Appendix B2) does not require the empirically-derived complexities that characterize the macrophyte model; there is also no need to distinguish among types of biomass. At present, CLEAN incorporates two functional groups of phytoplankton separated on the basis of size. The assumption is that nutrient uptake rates, sinking rates, and susceptibility to grazing are governed in large part by volume and correlative surface area. The model formulation is general, permitting redefinition of the present two groups through parameter changes; inclusion of additional groups is also easily accomplished because of the program structure.

Phytoplankton photosynthesis

The maximum photosynthesis rate, Pmax, is altered by reduction factors to account for nutrient limitation, and non-optimal light and temperature (Equation 2.2). Nutrient and light limitation is expressed by a construct:

$$\mu_t = 4 \bigg/ \left(\frac{1}{\mu_I} + \frac{1}{\mu_n} + \frac{1}{\mu_p} + \frac{1}{\mu_c} \right)$$

or generally:

$$\frac{1}{\mu_t} = \frac{1}{N} \sum_{i=1}^{N} \frac{1}{\mu_i}$$

in which N is used to normalize the function. Other formulations that have been tried are:

$$\mu_t = \min (\mu_I, \mu_n, \mu_p, \mu_c)$$

and

$$\mu_t = \mu_I * \mu_n * \mu_p * \mu_c$$

The first expression seems to be suitable for pure cultures but is not appropriate for natural assemblages with varying adaptions. The second expression is extreme in that even under near-optimal conditions the factors have fractional values which, when multiplied, can produce a severe reduction effect in the simulation.

The term for light limitation (Equation 2.21a) permits photosynthesis to increase exponentially to a maximum at the saturated light intensity, beyond which photosynthesis decreases with increased radiation. Mathematically, this relationship is:

$$\mu = Pmax \frac{I_o}{I_s} e^{\left(1 - \frac{I_o}{I_s} \right)}$$

and the normalized term is:

$$\mu_I = \frac{\mu}{Pmax} = \frac{I_o}{I_s} e^{\left(1 - \frac{I_o}{I_s} \right)}$$

Integrating over depth and time, the daily-averaged limitation for the given depth is as shown in Equation 1.1. The nutrient limitation terms represent saturation relationships analogous to Michaelis-Menten kinetics (Equation 2.21b).

Phytoplankton respiration

The respiration rate is a fraction of the maximum photosynthesis rate at optimal temperature multiplied by a reduction factor for non-optimal temperature (Equation 2.5).

Phytoplankton excretion

The excretion rate is considered to be a fraction of net photosynthate production as long as net production is positive (Equation 2.3).

Phytoplankton mortality

Mortality includes both grazing by herbivorous organisms (see "Consumers") and non-grazing mortality (Equation 2.4). The second factor is due primarily to sinking and is proportional to water temperature, which controls the specific gravity and viscosity of water.

Consumers

At present CLEAN has provisions for modeling seven consumers. These include three types of zooplankton: Herbivorous copepods and cladocerans and predatory copepods; three types of fish: A generalized predator which feeds on zooplankton and benthos, a scavenger and a piscivore; and benthic insect larvae, specifically chironomids. Traditionally, consumer modeling in the Biome has utilized a mass-balance equation with terms for consumption, predation, respiration, excretion, defecation, gonad production and non-predatory mortality (for example, Equation 4.1).

Recently it has been found advisable to develop a numbers-biomass model in order to simulate the complexities of consumer population and growth dynamics. This permits the calculation of mean weight which is necessary for the evaluation and symbolic switching of physiologic and ecologic processes. Such a model is particularly useful in simulating benthic insect larvae that emerge from the lake when their weight reaches a critical level (Appendix B6). However, the writers are presently implementing the model for fish in order to represent more realistically the significant changes in feeding and predation that accompany maturation.

Conceptually, one can divide consumer processes into two groups: Those that represent an overall change in biomass for the entire population (consumption, respiration, defecation, excretion and gamete production) and those that affect the numbers of individuals (recruitment, predatory and non-predatory mortality, promotion to next size class, emergence and emigration). Thus, the numbers equation has been formulated as

$$\frac{dN_j}{dt} = \frac{I_j}{W_k} - \frac{P_j}{Y_k} - (M_j + \sum_k C_{jk})/\overline{w}$$

in which

j	designates j^{th} size class,
I =	recruitment,
W_k =	weight of entering organisms,
P =	promotion (including emergence and emigration),
Y_k =	weight of organisms when they leave,
$\overline{w}$ =	mean weight of group at time t,
M =	non-predatory mortality, and
C =	predatory mortality (see also Equation 6.2).

Consumption

Consumption, or feeding, is a complex process which constitutes an extremely important linkage in the ecosystem; therefore, the formulation of the feeding term has received, and is continuing to receive, careful attention by both the aquatic and terrestrial Biome modeling groups. Multiple food sources, nonlinear physiologic effects and behavioral characteristics are all incorporated in the term (Appendix B3).

The adult feeding rate, C_{max}, is a complex function of temperature (Equation 3.2) so that at non-optimal

temperatures C_{max} is reduced accordingly. The process-temperature function:

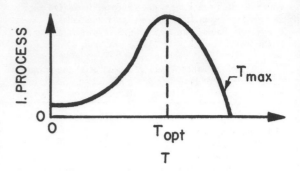

which occurs in the other process terms as well, permits an exponential increase in rate (the Q_{10} value) until the potential maximum is reached at the optimal temperature; beyond the optimal temperature the rate decreases rapidly until the lethal temperature is reached.

Behavioral corrections (Equation 3.11) are specific to the functional-group models. For the fish a density-dependent construct is used (cf. Equation 5.2). Benthic insect feeding is moderated by both dissolved oxygen levels and density of individuals (cf. Equation 6.12). At present, no behavioral correction is used for zooplankton.

The feeding rate is also dependent on a complex nonlinear relationship of consumer biomass and food supply, modified by a food preference or capturability factor (w_{ij}) to differentiate among food types. Equation 3.1 contains the construct used until recently by all the Biome modeling groups (Park et al., 1973). The writers are presently using a somewhat different formulation where w_i (B_i - $Bmin_i$) is the weighted biomass on which a consumer can feed.

$$\frac{w_{ij} (B_i - Bmin_i)}{\sum w_{ij} (B_i - Bmin_i)}$$

represents the preference for one food type compared to all food types. Minimum feeding area is represented by Q_j (which includes depth, z, to remove site-specificity). Density interference in feeding is represented by $r_j B_j$. Summing over i food sources, the term becomes

$$\frac{\sum_i w_{ij} (B_i - Bmin_i)}{(Q_j + r_j B_j) + \sum_i w_{ij} (B_i - Bmin_i)}$$

in which $(Q_j + r_j B_j)$ can be considered as a half-saturation "constant." The relationship between this

144

factor and consumer biomass can be shown graphically as:

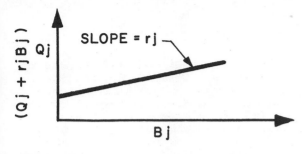

Juveniles generally feed at a higher rate than adults. Because C_{max} is defined as the maximum adult feeding rate, a population age-structure factor, b_j, is used to increase the feeding rate for a juvenile population (Equation 3.11). As the population reaches the carrying capacity, it is assumed that reproduction is greatly reduced and the population consists of adults; therefore, the age-structure correction factor is switched off. However, it is difficult to define the carrying capacity for many populations, and size-selective predation and other effects can alter the population structure greatly. Realization of this has given impetus to our development of a numbers-biomass model that can predict mean weight and permit differentiation of population size classes. In the numbers-biomass model the age-structure factor is replaced by an allometric correction factor utilizing the mean weight.

Respiration

Loss due to respiration is proportional to consumer biomass, with adjustments for temperature and behavior as discussed in "Consumption." Note the effect that crowding has on respiration in fish (Equation 5.2) which is not exhibited by zooplankton (Equation 4.3). The metabolic cost of food digestion and utilization is also included in the fish model (Equation 5.2). The effect that low oxygen levels has on respiration in benthic insects is portrayed realistically, too.

Mortality

The instantaneous rate of non-predatory mortality is corrected for behavior, crowding, age-structure and temperature as discussed in "Consumption" and "Respiration." Crowding is a mortality factor even for zooplankton (Equation 4.5).

Predatory mortality is treated the same as consumption, forming the linkage to higher trophic levels. In other words, it is the sum of the weighted feeding of all predators having non-zero food preference factors for the given consumer. In the fish model provision is also made for mortality due to fishing (Equation 5.7).

Defecation

Defecation, or egestion, is the elimination of that proportion of food that is not able to be assimilated. The proportion varies greatly depending on the type of food. Because many aquatic animals, being filter- and deposit-feeders, process food according to size and not quality, this is an important term.

Excretion

The elimination of metabolic wastes is expressed as a fraction of assimilated food; it is specific for the type of organism but is not specific for the type of food (Equation 5.4).

Exuvial loss

Arthropods exhibit a reduction in biomass due to loss of the exoskeleton at time of molting. At present this process is included in the benthic insect model (Equation 6.1) but it is not implemented in the zooplankton model. Expression of this process as a pulse is facilitated by the numbers-biomass model.

Gamete loss

During spawning a fraction of the biomass of the population is lost in the form of gametes. The rate of loss is determined by the fraction of biomass in gametes at time of spawning and the mortality rate for gametes; the term is switched on and off by minimum and maximum temperature and is modified by age-structure and behavior. The term is implemented in the fish model (Equation 5.5).

Promotion and emergence

In the benthic insect model the emergence term is taken to be $e_b w_c$, in which w_c is the weight of the organism when it emerges, e_b is zero when the mean weight is less than the threshold weight for emergence, and is a function of numbers and of the exponent of the difference between w_c and the current mean weight when the mean weight is equal to or greater than the threshold weight, w_s (Equation 6.12).

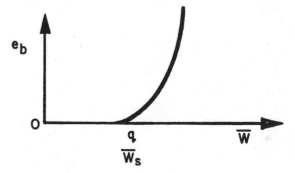

145

Promotion from one size class to another is handled in a like manner. Similar terms can be incorporated into other consumer models to represent maturation and size-dependent emigration.

Recruitment

The influx of benthic insect eggs is treated separately because the adult insects actually leave the ecosystem. The calculation of recruitment is based on the average weight and viability of eggs and the number of successfully emerging adults. Egg viability is reduced accordingly if the remaining benthic larval population is large, because of cannibalism. The adult to egg transfer function can also be modified to include the deleterious effect of wind if experience shows this to be important.

Decomposition and organic matter

Decomposition can be considered as two steps: 1) Hydrolysis of particulate organic matter (POM) to produce dissolved organic matter (DOM) and 2) uptake of this DOM by the decomposers (fungi, bacteria, and protozoans). The decomposers in turn produce inorganic carbon, nitrogen, phosphorus, and refractory dissolved organic matter. The decomposer-particulate matter aggregate serves as a food source for zooplankton, benthic insect larvae, and bottom-feeding fish.

Particulate organic matter

Inputs to the suspended and sedimented POM pools include allochthonous material, plant and animal wastes through mortality, defecation, leaf sloughing and exuvial loss, and resuspension of bottom sediments (Equation 7.2). Allochthonous material is treated as a driving variable. Loss terms include hydrolysis and ingestion by consumers.

Sedimentation and resuspension is a function of the potential for vertical mixing, as indicated by the thermal gradient, and wave agitation (Equation 7.8). Wave agitation is a result of wind stress and can be represented as such in the model. Alternatively, it is possible to use the change in barometric pressure to predict the depth of sediment at a given site (Fox and Davis, 1972); this construct is shown in Equation 7.8.

Hydrolysis of POM is a function of water temperature, pH, dissolved oxygen, biomass of decomposers and the amount of POM (Equation 7.7). The maximum hydrolysis rate is reduced by non-primal levels of temperature (see Equation 3.2), dissolved oxygen (Equation 7.41), and pH (Equation 7.71). The dissolved oxygen term has provision for a change in activity under anaerobic conditions (Equation 7.41). When decomposers are abundant the hydrolysis flux is proportional to the POM concentration modified by a surface area coefficient (Equation 7.7); when POM is abundant the flux is proportional to the biomass of decomposers; when both POM and decomposers are at intermediate levels the flux is proportional to both.

Dissolved organic matter

Inputs to the DOM pools in the water column and sediments include allochthonous loading (a driving variable), excretion, mortality, hydrolysis and diffusion across the sediment-water interface (Equation 7.3). Loss is due to uptake by decomposers (Equation 7.4) and is a complex function, with maximum uptake reduced by non-optimum levels of temperature and dissolved oxygen; it also includes a construct for saturation kinetics, with a minimum DOM level below which uptake is negligible.

Decomposers

Decomposer growth (Equation 7.1) is a function of the uptake of DOM (Equation 7.4), with loss terms for respiration, excretion, sedimentation, non-predatory mortality and ingestion by consumers. Respiration losses are considered proportional to decomposer biomass modified by temperature and oxygen levels (Equation 7.5). Excretion is a fraction of the material assimilated so long as uptake exceeds respiration; the fraction varies depending on the organic compound (Equation 7.6). Non-predatory mortality, which includes lysis and inactivation, is a fraction of the dissolved oxygen concentration (Equation 7.10). Mortality due to grazing is represented by a form of the general feeding term (Equation 3.1) for combined decomposers and particulate substrate.

Hydrologic balance

The hydrology of a drainage basin serves to couple the lake to the terrestrial ecosystem and, through outflow, may also significantly affect the concentrations of nutrients, suspended organic matter and plankton. The Hydrologic Transport Model, developed by the Lake Wingra modeling group, provides the means for predicting the effects of land use on nutrient influx; HTM can be run to obtain nutrient loadings to drive CLEAN, but there seems to be little reason to combine the models because of the lack of feedback.

The lake water balance is more intimately coupled to the dynamics of the ecosystem because of the effects of wash-out. The prediction of outflow rates is important to understanding the process of eutrophication in reservoirs and in many natural lakes. Outflow is of little concern in Lake George, where the mean residence time for water is ten years; but it does have an effect on Lake Wingra and should be included in any generalized model. The lake water balance model (Dettmann and Huff, 1972) given in Appendix B8 is presently being incorporated into CLEAN.

Lake circulation

Circulation can be subdivided into two components: Vertical mixing and horizontal transport. A simplified construct for vertical mixing due to wave agitation is incorporated in the decomposition model (Equation 7.8) as discussed under "Particulate Organic Matter." This construct also is being implemented in the model to predict sinking rates more accurately.

Horizontal circulation is of principal interest in ecosystem modeling because of the transport of nutrients, plankton and suspended organic matter from one area to another. Physical limnologic models have already reached a high degree of sophistication; two- and three-dimensional models of varying levels of revolution are available—some with unreasonable data requirements and others with requirements and output that are consistent with the resources and objectives of IBP studies.

Simulation and Evaluation

A purpose basic to the development of CLEAN is the forecasting of lake eutrophication. At present the model does not adequately represent the feedback between biotic components and nutrient pools, and further linkage is required for realistic simulation studies. However, inspection of the results of a simulation, using Lake Wingra data, shows that the patterns of predicted values are representative of many temperate-latitude, eutrophic lakes (Figure 3).

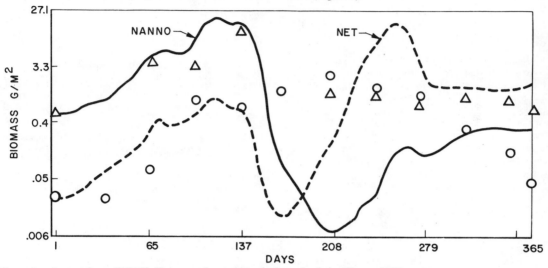

Figure 3a. Output from CLEAN: Nanno- and net phytoplankton for Lake Wingra, 1971.

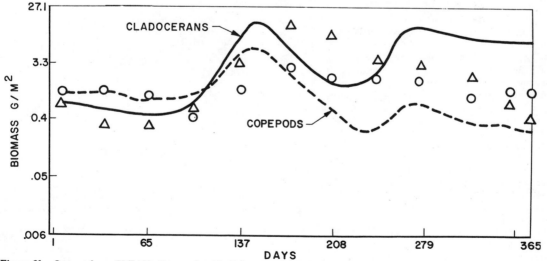

Figure 3b. Output from CLEAN: Copepods and cladocerans for Lake Wingra, 1971.

According to the assumptions given in the model, nanno-phytoplankton increase during the spring as light ceases to be limiting and temperature increases. The spring increase is followed by a decline produced by nutrient depletion, intensified grazing and increased sinking rates. Zooplankton, closely coupled to the algae by means of the feeding term, exhibit a similar increase and subsequent decline with a realistic lag in response time because of a slower growth rate.

In the summer net phytoplankton dominate over the nanno-phytoplankton in the simulation because of a lower half-saturation constant for the orthophosphate, a slower sinking rate, lower susceptibility to grazing, and higher optimal temperature. The zooplankton again increase in response to this new, though less preferable, food supply. In the fall the nanno-phytoplankton population recovers as lower water temperatures decrease the grazing and sinking rates.

Because this simulation utilized the same set of time-series data that were used to calibrate the model, it is not an adequate test of the model. The model contains so many parameters subject to tuning that it is possible to obtain a good simulation for the wrong reasons. However, from the inception of the Biome study there has been well defined strategy for evaluating the resulting models. Lake George and Lake Wingra fall at opposite ends of several environmental continua, including nutrient enrichment and morphometric characteristics. By changing site constants and substituting the appropriate driving variable values, CLEAN can be run for either lake (as well as for other lakes for which there are suitable data). Therefore, a model calibrated for Lake Wingra can be evaluated for an entirely different set of conditions at Lake George. Furthermore, because of the heterogeneity exhibited by Lake George (Park and Wilkinson, 1971), it is possible to evaluate models under a wide range of nutrient loadings in the one contiguous body of water.

Simplification for Management Purposes

As Levins (1968) has pointed out, a mathematical model may have theoretical realism, accuracy of prediction, and generality. Rarely does one encounter a model that possesses more than two of these attributes to any degree. In its present form, CLEAN exhibits both generality and ecologic realism; it has yet to be demonstrated that the accuracy of prediction is sufficient for it to be used as a management tool, especially in light of the cost of running such a complex model. However, the results of these simulations which are actually relatively inexpensive when compared with laboratory experimentation, give much insight into the apparent functionality of the ecosystem. Using the understanding gained from the basic research, simpler dynamic models can be produced for routine application in environmental management. One such model that has been developed in the Biome is WNGRA2. Another that has been strongly influenced by

CLEAN is SIMPLE, which was designed specifically to simulate conditions in Lake George but, with only slight modifications, can be applied to any lake (DiCesare and Bloomfield, 1973).

SIMPLE is an eighth-order nonlinear compartmental model (Figure 4) structured primarily to examine the effects of excessive phosphate loadings. Deriving variables are solar radiation, water temperature, allochthonous organic matter and phosphate loading. The state variables represent larger ecologic groupings than in CLEAN, the time step is weekly, and the model considers only the most significant transfers. However, the results are quite encouraging (Figure 5). Furthermore, the original version of SIMPLE was developed in approximately two weeks by a small class of graduate students—using DYNAMO and a basic understanding of CLEAN.

Selected Bibliography

Adams, M. S., M. D. McCracken, J.E. Titus, and W.H. Stone. 1972. Depth distribution of phytosynthetic capacity in a *Myriophyllum spicatum* L. community in Lake Wingra. EDFB Memo Rpt. 72-127. 9 p.

Allen, T. F. H., and J. F. Koonce. 1972. Multivariate approaches to algal strategies and tactics in the systems analysis of phytoplankton. EDFB Memo Rpt. 72-24. 53 p.

Dettmann, E.H. 1973. A model of seasonal changes in total nitrogen concentration in the Lake Wingra water column. EDFB-IBP-72-12. Oak Ridge National Laboratory, Oak Ridge, Tennessee. 36 p.

Dettmann, E.H., and D. Huff. 1972. A lake water balance model. EDFB Memo Rpt. 72-126. 22 p.

DiCesare, F., and J.A. Bloomfield. 1973. SIMPLE: A simplified ecosystem model for Lake George, New York. Manuscript in preparation.

Edmondson, W.T. 1960. Reproductive rates of rotifers in natural populations. Mem. 1st Ital. Idrobiol. 12:21-77.

Fisher, J., R.C. Kohberger, and J.W. Wilkinson. 1972. FIND—Freshwater Institute numeric data base. EDFB Memo Rpt. .72-61. 12 p.

Fox, W.T., and R.A. Davis, Jr. 1971. Computer simulation model of coastal processes in Eastern Lake Michigan, Technical Rept. 5 of ONR Task No. 388-092/10-18-68(414). Williams College. 114 p.

Gear, C.W. 1971. The automatic integration of ordinary differential equations. Comm. ACM, 14:176-179. March.

Hammerling, F.D. 1970. Kutta. *In* Westley, G. W., and J. A. Watts (eds). The computing technology center numerical analysis library. Union Carbide Corporation, Oak Ridge, Tennessee.

Hoopes, J., D. Patterson, M. Woloshuk, P. Monkmeyer, and T. Green. 1972. Investigations of circulation, temperature, and material transport and exchange in Lake Wingra. EDFB Memo Rpt. 72-117. 8 p.

Huff, D. D. 1971. Hydrologic transport of materials in ecosystems. EDFB Memo Rpt. 71-44. 7 p.

Huff, D. D. 1972a. Hydrologic transport of materials in ecosystems: Annual report for 1971-1972. EDFB Memo Rpt. 72-121. 12 p.

Huff, D.D. 1972b. HTM program elements, control cards, input data cards. EDFB Memo Rpt. 72-13. 46 p.

Ivarson, W.R., J.E. Jacques, and D.D. Huff. 1972. Report on implementation of lake and reservoir flow routing into the HTM. EDFB Memo Rpt. 72-135. 31 p.

Jacques, J. E., and D.D. Huff. 1972a. Open channel flow simulation with the hydrologic transport model. EDFB Memo Rpt. 72-134. 19 p.

Jacques, J.E., and D.D. Huff. 1972b. Snow accumulation and melt simulation. EDFB Memo Rpt. 72-136. 17 p.

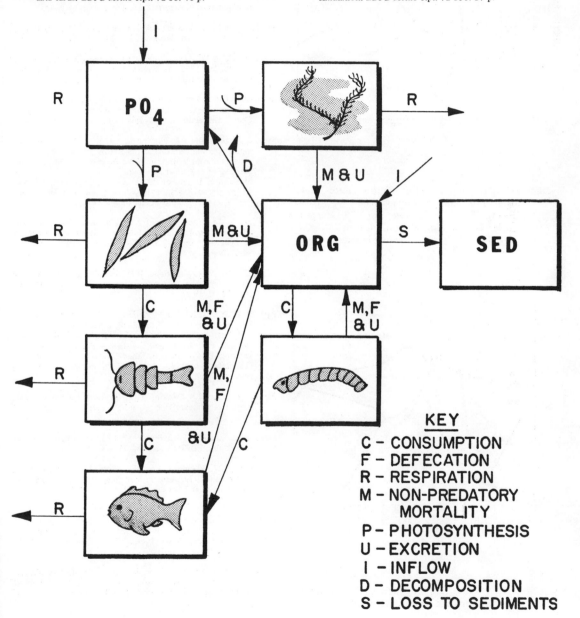

KEY

C – CONSUMPTION
F – DEFECATION
R – RESPIRATION
M – NON-PREDATORY MORTALITY
P – PHOTOSYNTHESIS
U – EXCRETION
I – INFLOW
D – DECOMPOSITION
S – LOSS TO SEDIMENTS

Figure 4. Flowchart of SIMPLE.

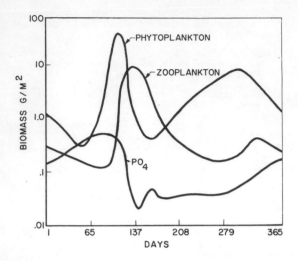

Figure 5a. Output from SIMPLE: Phytoplankton, zooplankton and orthophosphate for Lake George, Sta. 1, 1972. Note: For Lake George 1 G/M^2 is approximately equivalent to 25μ gC/l (biotic) and 50μ gP/l (orthophosphate).

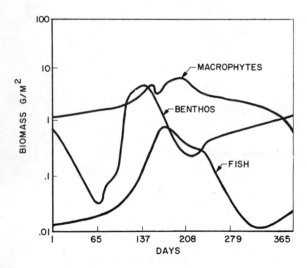

Figure 5b. Output from SIMPLE: Macrophytes, benthos and fish for Lake George, Sta. 1, 1972.

Kitchell, J.R., J.F. Koonce, R.V. O'Neill, H.H. Shugart, J.J. Magnuson, R.S. Booth. 1972. Implementation of a predator-prey biomass model for fishes. EDFB Memo Rpt. 72-118. 57 p.

Koonce, J.F., and A.D. Hasler. 1972. Phytoplankton succession and a dynamic model of algal growth and nutrient uptake. EDFB Memo Rpt. 72-114. 112 p.

Levins, R. 1968. Evolution in changing environments. Monographs in Population Biology II. Princeton University Press.

Lytle, R., and N.L. Clesceri. 1972. Lake George, N.Y. Hydrology, 1971-72. EDFB Memo Rpt. 72-65. 4 p.

MacCormick, A.J.A., O.L. Loucks, J.F. Koonce, J.F. Kitchell, and P.R. Weiler. 1972. An ecosystem model for the pelagic zone of Lake Wingra. EDFB Memo Rpt. 72-122. 103 p.

McAllister, C.D. 1970. Zooplankton rations, phytoplankton mortality and the estimation of marine production. Marine Food Chem., University of California Press.

McNaught, D.C., and J.A. Bloomfield. 1972. A resource allocation model for herbivorous zooplankton predicts community changes during eutrophication. Paper presented at Annual Meeting of the American Association for the Advancement of Science. December 26-31, AAAS Symposium, Advances in Integrated Research–Experimentation and Modeling in the Eastern Deciduous Forest Biome, IBP.

Nagy, J., R.C. Kohberger, and J.W. Wilkinson. 1972. ADLIB–Abstract Data LIBrarian, a bibliographic retrieval system for data set abstracts with the FIND System. EDFB Memo Rpt. 72-62. 8 p.

O'Neill, R.V., R.A. Goldstein, H.H. Shugart, and J.B. Mankin. 1972. Terrestrial ecosystem energy model. EDFB Memo Rpt. 72-19. 39 p.

Park, R.A., R.V. O'Neill, J.A. Bloomfield, H.H. Shugart, Jr., R.S. Booth, J.R. Koonce, M.S. Adams, L.S. Clesceri, E.M. Colon, E.H. Dettmann, J.A. Hoopes, D.D. Huff, Samuel Katz, J.F. Kitchell, R.C. Kohberger, E.J. LaRoy, D.C. McNaught, J.L. Petersen, Don Scavia, J.E. Titus, P.R. Weiler, J.W. Wilkinson, C.S. Zahorcak. 1973. A generalized model for simulating ecosystems. Simulation *in review*.

Park, R.A., and J.W. Wilkinson. 1970. Lake George modeling project preliminary progress report. EDFB Memo Rpt. 70-2. 93 p.

Park, R.A., and J.W. Wilkinson. 1971. Lake George modeling philosophy. EDFB Memo Rpt. 71-9. 62 p.

Park, R.A., J.W. Wilkinson, J.A. Bloomfield, R.C. Kohberger, and C. Sterling. 1972. Aquatic modeling, data analysis, and data management at Lake George, New York. EDFB Memo Rpt. 72-70. 32 p.

Petersen, J.L., and W.L. Hilsenhof. 1972. Preliminary modeling parameters for aquatic insects and sanding crop comparisons with other Lake Wingra biota. EDFB Memo Rpt. 72-116. 7 p.

Titus, J.E., M.S. Adams, P.R. Weiler, R.V. O'Neill, H.H. Shugart, R.S. Booth, and R.A. Goldstein. 1972. Production model for *Myriophyllum spicatum* L. EDFB Memo Rpt. 72-108. 17 p.

Appendix A

Implementation of CLEAN

CLEAN has been implemented in real-time mode to take full advantage of man-machine interaction. FIND, the companion data handling system (Wilkinson et al., 1972), also is accessed in real time.

The main routine handles all input and output and calls subroutines for integrating, saving updated parameter values and displaying results in a table or plot. A modification of the Runge-Kutta-Merson integration algorithm (Hammerling, 1970), is presently used to solve the state-variable equations. It is a fourth order variable-step method which operates at a user-specified maximum local error. Implementation of a variable order predictor-correction method (Gear, 1971) is expected to decrease program execution time significantly. All arithmetic operations are carried out in extended precision to minimize machine-introduced error.

Each command has the form:

*xxx

in which xxx is a mnemonic for the command name. In addition to eleven commands, in-line parameter editing is also available. Parameter changes may be entered in a free-form format. Parameter values can be saved from one user session to the next by means of a file restore (*RES).

Mnemonic	Syntax	Command Usage
PRI	1) *PRI	To print initial conditions
PRP	1) *PRP 2) nn (Parameter list code)	To print a specific list of parameters
EDI	1) *EDI 2) b & BEGIN (initial condition list) & END	To edit initial conditions
EDP	1) *EDP 2) b & EDIT (parameter list) & END	To edit parameters
(Inline editing)	1) X = nnn, Y=jjj, ... where X & Y are variables and nnn and jjj are their new values	To directly edit initial conditions and parameters
LAK	1) *LAK 2) nnn (Lake code)	To run simulation for the lake specified
INT	1) *INT 2) KILL - to abort integration after a user specified interval	To solve system of differential equations and tabulate results; the "KILL" subcommand is used to abort meaningless runs.
TAB	1) *TAB	To tabulate integration results
PLT	1) *PLT	To plot integration results
MSG	1) *MSG	To retrieve messages for user
END	1) *END	To normally terminate program
RES	1) *RES	To normally terminate program and save editing changes to parameters and initial conditions until next session
(Line delete)	Site Dependent	To delete an incorrect input line
(Break interrupt)	Site Dependent	To terminate program while it is generating output
(Execute)	Site Dependent	To start program

Examples of each command are given below; program response is indented.

***** CLEAN V1.0 *****
FWI TIME/SHARE MODELS
LAST UPDATE 8/28/73
FOR NEW MSGS DO A *MSG
READY
*MSG
  NEW MSGS

8/28/73 JSF IS WORKING ON GEAR'S METHOD
08/28/73 DON S. IS ACTIVATING PREDATORY ZOO.
8/28/73 TO RESTART MODEL AFTER A FAIL, SAY
      XIT ACLEAN
8/15/73 REPORT ANY ODD ERRORS TO D. SCAVIA
READY
*PRP
SECT: 01-15
01
CMAX = 2.00 .50 .05 .25 Q = 3.2 1.0 4.0 4.0 R = .0 .0 .0
    .0
MORE? (YES/NO)
NO
READY

*PRI
BIOMIC = .600+000 .200-001 .750+000 .135+001
TFIRST = 1.00 TLAST = 365.00 STEP = 14.000
      ACCURG = .500-002 IMAX = 18
READY

*EDP
ENTER &EDIT...PARMS...&END.
&EDIT CMAX(3)=1. &END
READY

*EDI
ENTER &BEGIN ...INITIA CONDITIONS...&END.
&BEGIN BIOMIC(1)=.65 &END
READY

CMAX(3)=2.,BIOMIC(1)=.6
READY

TIME	NANO	NET	CLAD	COPE	FISH	PREDZ	DET	BENTHOS
1.0	.600	.020	.750	1.350	1.50	1.00	8.53	.01
15.0	.612	.033	.632	1.272	1.55	1.10	11.45	.01
29.0	.701	.063	.542	1.197	1.60	1.30	9.37	.01
43.0	.839	.124	.456	1.120	1.65	1.50	9.35	.01
57.0	1.087	.252	.377	1.045	1.72	1.60	13.55	.01
71.0	1.535	.489	.315	.991	1.30	1.75	25.42	.01
85.0	.761	.307	.248	.935	1.90	1.90	22.22	.01
99.0	.274	.104	.122	.698	1.95	2.10	21.05	.22
113.0	.734	.114	.046	.461	2.00	2.20	23.64	.95
127.0	1.112	.061	.017	.311	2.05	2.00	32.55	1.15
141.0	17.933	.170	.053	.627	2.10	.60	14.96	.60
155.0	.348	.004	9.647	5.814	2.13	.46	12.10	.25
169.0	.000	.000	1.280	2.422	2.25	.80	31.93	1.04
183.0	.001	.002	.126	.724	2.40	4.75	32.10	.32
197.0	.037	.098	.011	.170	2.75	4.56	31.04	.42
211.0	.498	2.754	.002	.088	3.15	3.32	32.15	.30
225.0	4.114	20.163	.061	.759	3.90	1.80	33.70	.08
239.0	.098	.588	5.258	6.748	4.90	1.60	34.32	.01

STAR TIME	NANNO	NET	CLAD	COPE
1.00	6.000 -001	2.000 -002	7.500 -001	1.350+000
15.00	6.120 -001	3.330 -002	6.373 -001	1.272+000
29.00	7.010 -001	6.280 -002	5.422 -001	1.197+000
43.00	3.393 -001	1.838 -001	4.558 -001	1.120+000
57.00	1.087+000	2.524 -001	3.772 -001	1.045+000
71.00	1.535+000	4.890 -001	3.150 -001	9.912 -001
85.00	7.607 -001	3.067 -001	2.480 -001	9.348 -001
99.00	2.742 -001	1.038 -001	1.222 -001	6.978 -001
113.00	7.344 -001	1.141 -001	4.634 -002	4.608 -001
127.00	1.112+000	6.112 -002	1.690 -002	3.105 -001
141.00	1.793+000	1.704 -001	5.290 -002	6.267 -001
155.00	3.485 -001	3.785 -003	9.647+000	5.814+000
169.00	1.608 -004	1.624 -004	1.280+000	2.422+000
183.00	1.441 -003	1.754 -003	1.264 -001	7.238 -001
197.00	3.743 -002	9.830 -002	1.111 -002	1.699 -001
211.00	4.982 -001	2.754+000	2.003 -003	8.760 -002
225.00	4.114+000	2.016+001	6.036 -002	7.593 -001
239.00	9.762 -002	5.881 -001	5.258+000	6.748+000
READY				

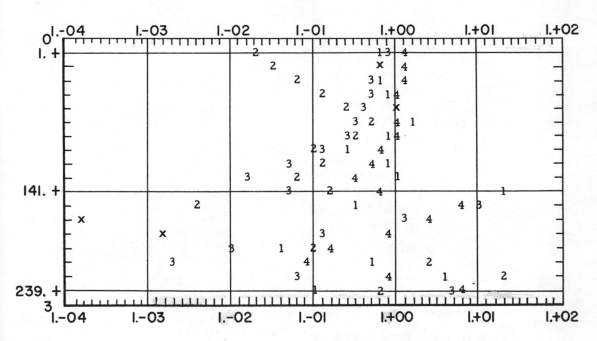

READY

*LAKE
INPUT LAKE CODE:
GEO
READY
*RES

FILES RESTORED. ELAPSED TIME FOR THIS RUN WAS

31855 MSEC. READY!

READY
READY!

153

Table 1. Macrophytes.

$$\text{(1.1)} \quad P_{net,m} = (1-X)\left[\frac{P_{max,m}}{\kappa_1}\right.$$

$$\ln\left(\frac{A + I_o \exp[-z(\epsilon+\kappa_2 B_p)]}{A + I_o \exp[-z(\epsilon+\kappa_2 B_p) - \kappa_1 F]}\right)$$

$$\left. \cdot f(T)g(V) - \rho_f F\right]$$

$P_{net,m}$ = net primary production (quantity of material fixed by photosynthesis and available for plant growth)

X = fraction of photosynthetic product excreted from plant and lost

$P_{max,m}$ = maximum photosynthetic rate (under conditions of optimal light, temperature and nutrients)

κ_1 = coefficient for light extinction due to plant leaves

A = light saturation coefficient (at light intensity A, photosynthesis rate = $1/2\ P_{max,m}$)

I_0 = light intensity at water surface driving or extrinsic variable which varies during the year)

z = water depth

ϵ = coefficient for light extinction due to water

κ_2 = coefficient for light extinction due to phytoplankton biomass

B_p = phytoplankton biomass

F = leaf area index (total surface of leaves suspended over a unit area of lake bottom)

V = water current velocity

$g(V)$ = function of water current velocity necessary to renew nutrients in boundary zone of plant; ≤ 1.0

T = temperature, $^\circ$C

$f(T)$ = function of temperature (see Table 3.) ≤ 1.0

ρ_f = respiration rate of plant leaves; a function of temperature

$$\text{(1.2)} \quad \frac{dL}{dt} = L^*\delta(t_p) + G_L - (R_L + M_L)$$

$$\delta = \begin{cases} 1, & t=t_p \\ 0, & \text{otherwise} \end{cases} \quad \ldots \ldots \ldots .1.21$$

L = biomass of plant leaf tissue

t = time

t_p = time for initial plant growth in spring, based on cumulative effect of temperature during early spring

L^* = initial biomass pulse of arbitrarily small value

G_L = growth of new leaf tissue

R_L = leaf respiration

M_L = leaf mortality (sloughing of entire leaves)

Table 1. Continued

$$\text{(1.3)} \quad G_L = \lambda L \left(1 - \frac{L}{nF_{opt}}\right)\left(\frac{S-S_o}{S_1+S-S_o}\right)$$

$$\nu(nF_{opt}-L)\nu(S-S_o)$$

$$\nu(a) = \begin{cases} 0, & a \leq 0 \\ 1, & a > 0 \end{cases} \quad \ldots \ldots \ldots .1.31$$

λ = fractional increase in leaf biomass per unit time

n = units leaf biomass per unit leaf area

F_{opt} = optimal leaf area index

S = stored pool of carbohydrates resulting from photosynthesis

S_0 = minimal level of carbohydrates reserved for overwintering

S_1 = carbohydrate level at which growth is $1/2$ maximum

$$\text{(1.4)} \quad F_{opt} = \frac{1}{\kappa_1} D\nu(D)$$

$$D = \ln \frac{[P_{max}(1-\chi)q(V)f(T) - \rho_f]\exp[-z(\epsilon+\kappa_2 B_p)]}{A\rho_F}$$

$$\text{(1.5)} \quad R_L = \rho_L L\nu[\nu(S_o - S) + \nu(F - F_{opt})]$$

ρ_L = fraction of leaf biomass lost by respiration per unit time; a function of temperature

$$\text{(1.6)} \quad M_L = \gamma_1 L[\nu(t-t_1)\nu(t_2-t) + \nu(t-t_3)\nu(t_4-t)]$$

$$- \gamma_2 L\nu(F_{opt}-F)$$

γ_1 = fraction of leaf biomass sloughed per unit time during sloughing periods

γ_2 = fraction of leaf biomass sloughed per unit time when F_{opt} is exceeded by F

t_1 to $t_2|$ = initial period of sloughing

t_3 to t_4 = subsequent period of sloughing

$$\text{(1.7)} \quad \frac{dR}{dt} = G_R - R_R \nu(S_o-S)$$

R = biomass of roots

G_R = growth of root tissue

R_R = root respiration

$$1.8 \quad G_R = \mu R \left(1 - \frac{R}{R_{max}}\right)\left(\frac{S-S_o}{S_1+S-S_o}\right)\nu(S-S_o)$$

μ = fractional growth of root biomass per unit time

Table 1. Continued.

(1.9) $R_R = \rho_R R \nu (S_o - S)$

(1.10) $\dfrac{dS}{dt} = P_{net,m} - \rho_R R \nu (S - S_o) - \rho_L L \nu (S - S_o) \nu (F_{opt} - F)$

$\qquad\qquad - \alpha (G_L + G_R)$

α = inverse efficiency of conversion of labile carbohydrates to leaf and root biomass

Table 2. Phytoplankton.

(2.1) $\dfrac{dB_p}{dt} = (P_{net,p} - G_{p,z} - U_p - M_p) B_p$

$G_{p,z}$ = rate of grazing of zooplankton on phytoplankton
U_p = excretion rate
M_p = non-grazing mortality rate

(2.2) $P_{net,p} = (P_{max,p} n / [\sum_i 1/\mu_i] - R_p) f(T)$

$\mu_1 = \left[\dfrac{2.71828 \Phi_p}{I_s + \kappa_2 B_p} \right] \left[\exp\left(-\dfrac{I_o}{I_s \Phi_p} \exp[-z(+\kappa_2 B_p)] \right) \right.$

$\qquad\qquad \left. - \exp\left(-\dfrac{I_o}{I_s \Phi_p} \right) \right] \quad \cdots \cdots \cdots \cdots \; 2.21a$

$\mu_i = \dfrac{C_i}{N_i + C_i} , \quad i = 2, 3, \ldots, n \quad \cdots \cdots \; 2.21b$

R_p = respiration
μ_i = ith limiting factor: μ_1 = light; μ_i, $i \neq 1$, = nutrients
C_i = concentration of nutrient i
N_i = concentration of nutrient i at which photosynthesis rate = $1/2 \; P_{max,p}$
I_s = saturation light intensity (light intensity above which photosynthesis is maximum)
ϕ_p = photoperiod (number of hours of light in a 24 hour period)

(2.3) $U_p = \begin{cases} a \, P_{net,p}, & P_{net,p} \geq 0 \\ \\ 0, & P_{net,p} < 0 \end{cases}$

a = fraction of net photosynthate lost per unit time

(2.4) $M_p = bT$

b = fraction of algal biomass dying or sinking to lake bottom per unit time and temperature

Table 2. Continued.

(2.5) $R_p = R_{max} V^X e^{X(1-V)}$

$R_{max} = K P_{max,p}$
see 3.2 for V and X

Table 3. Trophic interaction or feeding term.

(3.1) $C_{i,j} = C_{max_j} H_{j,C} \left(\dfrac{w_{ij} B_i B_j}{Q_j + r_j B_j + \sum_i w_{ij} B_i} \right)$

$H_{j,n} = a_{j,n} \left(b_{j,n} \dfrac{(K_j - B_j)}{K_j} + 1 \right) c_{j,n} \quad \cdots \; .3.11$

C_{ij} = consumption of ith prey by jth organism
Q_j, r_j = environmental and population interation coefficients
$w_{i,j}$ = coefficient relating preference, availability, capturability, etc., of i as a food for j
K_j = carrying capacity
$a_{j,n}$ = correction factor for behavioral effects on the nth process (e.g., n=C for consumption, R for respiration, M for mortality, F for fishing, G for gamet production) for the jth organism
$b_{j,n}$ = correction factor for effects of population age structure on nth process
$c_{j,n}$ = correction factor for physiological effects on nth process

(3.2) $c_{z,C} = f(T) = V^X \exp[X(1-V)]$ for zooplankton

$V = \dfrac{T_{max} - T}{T_{max} - T_{opt}} \quad \cdots \cdots \cdots \cdots \; .3.21$

$X = \dfrac{W^2 [1 + \sqrt{(1 + 40/W)}]^2}{400} , \quad \cdots \cdots \; .3.22$

$W = (\ln S_Q)(T_{max} - T_{opt}) \quad \cdots \cdots \cdots \; .3.23$

T_{max} = upper lethal temperature
T_{opt} = optimum temperature (temperature at which rate is maximum)
S_Q = Q_{10} value (factor in long range by which rate is increased given a $10^{\circ}C$ increase in temperature)
$(C_{j,k}$ = consumption of jth organism by kth predator)

Table 4. Zooplankton.

$$(4.1) \quad \frac{dB_z}{dt} = \sum_{i=1}^{n} C_{i,z} \sum_{k=1}^{m} c_{z,k} \, (R_z + U_z + F_z + M_z)$$

B_z = biomass of zooplankton
R_z = respiration rate
U_z = rate of metabolic excretion
F_z = rate of egestion
M_z = rate of non-predatory mortality
(with $a_{z,C}$, $b_{z,C}$, $c_{z,C}$ in $C_{i,z}$)

$$(4.2) \quad F_z = \sum_{i=1}^{n} f_{i,z} C_{i,z}$$

$f_{i,z}$ = fraction of food supply i consumed but not assimilated

$$(4.3) \quad R_z = R_{max} H_{z,R} B_z \text{ (with } a_{z,R}, b_{z,R}, c_{z,R} \text{ in } H_{z,R})$$

$$(4.4) \quad U_z = u_z \sum_i C_{i,z}$$

u_z = fraction of food assimilated and subsequently excreted

$$(4.5) \quad M_z = H_{z,M} B_z \left(\frac{d_{z,M} B_z}{K_j} + 1 \right)$$

$d_{z,M}$ = density-dependent term for increased mortality due to overcrowding

Table 5. Fish.

$$(5.1) \quad \frac{dB_f}{dt} = \sum_{i=1}^{n} C_{i,f} - \sum_{k=1}^{m} C_{f,k} - (F_f + R_f + U_f + G_f M_f + C_{f,man})$$

B_f = biomass of fish
R_f = respiration rate
U_f = metabolic excretion rate
G_f = rate of gamet production
M_f = non-predatory mortality rate
$C_{f,man}$ = mortality rate due to fishing
F_f = egestion rate

$$(5.2) \quad R_f = s \sum_i C_{i,f} + H_{f,R} B_f \left(1 + \frac{d_{f,R}}{K_f} B_f \right)$$

with $a_{f,R}$, $b_{f,R}$ $c_{f,R}$ in $H_{f,R}$

s = metabolic cost of food digestion and utilization
$d_{f,R}$ = density-dependent term for increased respiration due to overcrowding

Table 5. Continued.

$$(5.3) \quad F_f = \sum_i f_{i,f} C_{i,f}$$

$f_{i,f}$ = fraction of ith ingested food that is not assimilated

$$(5.4) \quad U_f = u_f \sum_i C_{i,f}$$

u_f = fraction of assimilated food that is excreted

$$(5.5) \quad G_f = H_{f,G} Z_G \kappa_G B_f \text{ with } a_{f,G}, b_{f,G}, c_{f,G} \text{ in } H_{f,G}$$

Z_G = instantaneous rate of gamet mortality
κ_G = fraction of adult biomass in gamets at spawning
$c_{f,G}$ = temperature switched permitting spawning between a maximum and minimum temperature

$$(5.6) \quad M_f = Z_M H_{f,M} B_f \left(\frac{d_{f,M} B_f}{K_f} + 1 \right)$$

with $a_{f,M}$, $b_{f,M}$, $c_{f,M}$ in $H_{f,M}$

Z_M = rate of adult natural mortality

$$(5.7) \quad C_{f,man} = Z_F H_{f,F} B_f \text{ with } a_{f,F}, b_{f,F}, c_{f,F}, \text{ in } H_{f,F}$$

Z_F = rate of adult mortality due to fishing

Table 6. Benthos.

$$(6.1) \quad \frac{dB_{b_j}}{dt} = \sum_i C_{i,b_j} - (\sum_i f_{i,b_j} C_{i,b_j}) - ($$

$$- (u_{b_j} \sum_i C_{i,b_j}) - (x_{b_j} B_{b_j})$$

$$+ I_{b_j} - P_{b_j} - M_{b_j} - \sum_k C_{b_j,k}$$

B_{b_j} = biomass of benthos in jth size class: j=1, for instars 1-3; j=2, for 4th instar
f_{i,b_j} = fraction of ith food not assimilated (i.e., egested)
ρ_{b_j} = fraction of biomass respired per unit time = f(T,DO,W), in which T=temperature, DO = dissolved oxygen near sediments, W = organism's weight
u_{b_j} = fraction of biomass lost by metabolic excretion
x_{b_j} = fraction of biomass lost by molting of old exoskeleton = f(W)

| Table 6. Continued. | Table 7. Decomposition. |

Table 6. Continued.

I_{b_j} = influx into class. For j=1, = egglaying = iW_e; for j=2, maturation of 3rd instars into 4th: $p_{b_1} W_3$.

i = number of eggs that hatch = $f(N_1, N_2)$

W_e = average weight of an egg

p_{b_1} = number of molting 3rd instars (maturing into 4th)

W_3 = the critical weight of a 3rd instar at which molting is induced

P_{b_j} = promotion out of class. For j=1, = maturation of 3rd instar into 4th (= I_{b_2}). For j=2, = emergence of 4th instars to adults, = $e_b W_c$.

e_b = number of emerging insects

$$= \begin{cases} 0 & , \overline{W} < \overline{W}_s \\ N_{b_2} \kappa_1 (\exp[\kappa_2(\overline{W} - W_c)] - 1), & \overline{W} \geq \overline{W}_s \end{cases} \quad .6.11$$

W = mean weight of population

W_s = critical population mean weight to begin emergence

W_c = actual organism's weight to induce emergence

κ_1, κ_2 = fitted coefficients

$$M_{b_j} = Z_{b_j} H_{b_j, M} \left(1 + \frac{H_{b_j, M} B_{b_j}}{K_{b_j}}\right) B_{b_j} \quad .6.12$$

M_{b_j} = non-predatory mortality

$$H_{b_j, M} \begin{cases} a_{b_j, n} = & \text{physiological effects due to DO for } n=M,C \\ b_{b_j, n} = & \text{correction term for effect of mean weight on process} \\ & \left(no \left(\frac{K_{b_j} - B_{b_j}}{K_j} + 1\right) term\right) \\ = & \alpha \overline{W}^{\beta-1} \end{cases}$$

$c_{b_j, n}$ = physiological effects due to T for n=M,C

Z_{b_j} = instantaneous rate of non-predatory mortality

$d_{b_j, M}$ = increase in mortality due to density-dependent factors (overcrowding)

K_{b_j} = carrying capacity

α, β = fitted coefficients

$$(6.2) \quad \frac{dN_{b_j}}{dt} = \frac{1}{\overline{W}_{b_j}} (-M_{b_j} - \sum_k C_{b_j, k}) + \frac{I_{b_j}}{W_k} - \frac{P_{b_j}}{Y_k}$$

N_{b_j} = numbers of jth class of benthos

W_{b_j} = mean weight of jth class

$W_k = \begin{cases} W_e, k=1 \\ W_3, k=2 \end{cases}$

$Y_k = \begin{cases} W_3, k=1 \\ W_c, k=2 \end{cases}$

Table 7. Decomposition.

$$(7.1) \quad \frac{dB_{d_i}}{dt} = V_{d_i} - (R_{d_i} + U_{d_i} + S_{d_i} + M_{d_i} + \sum_i C_{d_i B_j})$$

B_{d_i} = biomass of the ith decomposer group

V_{d_i} = rate of uptake of organics

R_{d_i} = respiration rate

U_{d_i} = rate of excretion of inorganics and refractory organics

S_{d_i} = rate of sedimentation (or resuspension)

M_{d_i} = rate of mortality due to lysis, inactivation or micropredators

$$(7.2) \quad \frac{dB_{P_i}}{dt} = I_{P_i} + \sum_j F_j + \sum_j K_{p, M_j} M_j + S_{P_i}$$
$$- (H_{P_i} + \sum_j C_{P_i, j})$$

B_{P_i} = mass of the ith Particulate Organic Matter compartment

I_{P_i} = external loading of POM

F_j = defecation rate of the jth consumer group

K_{p, M_j} = input rate of POM due to non-predatory mortality

S_{P_i} = rate of sedimentation (or resuspension)

H_{P_i} = hydrolysis rate

$C_{P_i, j}$ = rate of grazing by jth consumer group

$$(7.3) \quad \frac{dB_{DOM_i}}{dt} = I_{DOM_i} + \sum_{j=1}^{n} U_j + \sum_j K_{p, M_j} M_j + H_{P_i}$$
$$+ D_{DOM_i} - V_d.$$

B_{DOM_i} = mass of the ith Dissolved Organic Matter compartment

I_{DOM_i} = external loading of ith DOM

U_j = excretion rate of organics by the jth biotic group

K_{p, M_j} = input rate of DOM due to non-prediatory mortality

D_{DOM_i} = diffusion rate of ith organic between sediment and water

$$(7.4) \quad V_{d_i} = V_{max, d_i} a_{d_i, V} c_{d_i, V}$$

$$\left(\frac{B_{DOM_j} - DOM_{min_j}}{K_{DOM_i} + B_{DOM_j} - DOM_{min_j}}\right) B_{d_i}$$

$$\cdot \nu (B_{DOM_j} - DOM_{min_i})$$

Table 7. Continued.

in which i=ith decomposer group, j=jth DOM group

V_{max,d_i} = maximum uptake rate

$c_{d_i,V}$ = effect of temperature on uptake (see Equation 3.2)

$a_{d_i,V}$ = effect of dissolved oxygen

$$= 1 \quad K_{V,O_2} \, \nu(O_{2,min} \quad O_2) \quad \ldots \ldots .7.41$$

DOM_{min_j} = DOM level below which uptake is negligible

X_{DOM_j} = half-saturation constant for uptake

$O_{2,min}$ = oxygen level for anaerobic uptake

K_{V,O_2} = change in uptake due to anaerobic conditions

$$(7.5) \quad R_{d_1} = R_{max,d_i} \, {}^c d_i,R \, {}^a d,R \, {}^B d_i$$

R_{max,d_i} = maximum respiration rate

$$(7.6) \quad U_{d_i} = \sum_j u_j \, \nu(O_{2,min} \quad O_2)(V_{d_i} - R_{d_i}) \, \nu \, (V_{d_i} - R_{d_i})$$

u_j = percent of net assimilation which is excreted for jth compound

$$(7.7) \quad H_{p_j} \quad H_{max,p_j} \, {}^c p_j,H \, {}^a p_j,H \, {}^h p_j,H$$

$$\left(\frac{B_{d_j}}{K_{sv_j} B_{p_j} + B_{d_i} + K_{POM_j}} \right) B_{p_j}$$

H_{max,p_j} = maximum hydrolysis rate

$h_{p_j,H}$ = effect of pH on hydrolysis

$$= \exp \left[- \left(\frac{pH - pH_{opt}}{\sigma} \right)^2 \right] \quad \ldots \ldots .7.71$$

K_{sv_j} = available surface area parameter

K_{POM_j} = saturation constant for hydrolysis

pH = pH of water or sediment interstitial water

pH_{opt} = optimum pH for hydrolysis

σ = pH "bandwidth" constant for hydrolysis

$$(7.8) \quad S_{d_1} = \psi_{d_i} \exp \left[K_{sed_{d_i}} \left(\frac{\Delta T}{\Delta z} \right) \right]^{\Delta_{P_{bar}} / \Delta t} \cdot B_{d_i}$$

$\Delta T / \Delta z$ = maximum vertical thermal gradient

Table 7. Continued.

$\Delta_{P_{bar}}$ = change in barometric pressure over past day

$K_{sed_{d_i}}$ = thermal gradient constant

ψ_{d_i} = sedimentation rate constant

$$(7.9) \quad D_{DOM_i} = K_{DIF} (B_{DOM_i} - B_{DOM_{i+1}})$$

K_{DIF} = boundary layer resistance coefficient

$$(7.10) \quad M_{d_i} = K_{d_i,M} \, {}^a d_i,M \, B_{d_i}$$

$K_{d_i,M}$ = non-macropredatory mortality

Table 8. Lake water balance.

$$(8.1) \quad \frac{dV}{dt} = (p-e) A(E) + Q_s f(E)$$

$$+ \sum_{i=1}^{n} \left[K_i \left(\frac{w_i - E}{D_i} \right) L_i (m_{0i} + \alpha(w_i - w_{0i})) \right]$$

$$- K_1 (E - E_c)^{3/2}$$

V = volume of lake water

$A(E)$ = lake surface area as a function of lake stage, E

p = precipitation rate

e = evaporation rate

$f(E)$ = elevation-dependent correction factor for inflow

Q_s = surface inflow rate

n = number of observation wells

K_i = saturated hydraulic conductivity at observation well i

w_i = elevation of groundwater table at well i

E = lake level

D_i = distance of well from lake

L_i = effective length of shoreline associated with area of well i

m_{0i} = base value for effective saturated thickness

α = parameter which characterizes dependence of M on w

w_{0i} = index elevation

K_i = coefficient for uncontrolled outflow

E_c = elevation of spillway

A PHYSICAL MODEL FOR SIMULATION OF AQUATIC ECOSYSTEMS

J. W. Falco and W. M. Sanders III [1]

Introduction

Physical modeling and simulation of aquatic eco-systems are relatively new developments in environmental science. Although work with microcosms has a long history and manipulations of partially controlled streams have been carried out for a number of years (Phinney and McIntire, 1965), the construction of totally controlled aquatic ecosystems which have the capability of mimicking complex natural ecosystems has begun only recently. The reason for delays between the development of partially controlled streams and studies with microcosms and development of scaled physical models is largely one of cost. The facility that is described in this paper was built at a cost of approximately $1 million. This cost, however, has also partially supported the testing and operation as well as the design and construction of an environmental chamber called the Aquatic Ecosystem Simulator (AEcoS).

The AEcoS was designed as a process physical model, as opposed to a geometric or hydrodynamic model, which would simulate biological and chemical responses of natural flowing streams to a wide range of environmental perturbations. In addition to providing information on microbial growth kinetics, AEcoS is to provide a capability for introducing significant material gradient along the length of a water channel and thus provide a facility to study the role of transport phenomena in environmental water quality problems. A third feature of the AEcoS is the variable light source which simulates solar radiation not only in terms of total intensity but also in terms of spectral distribution. Obviously such a light source is intended for the study of photosynthetic processes.

Description of Facility

Since the AEcoS facility has been described in detail in a previous publication (Sanders and Falco, 1973), only a brief description will be presented. The major portion of AEcoS (Figure 1) is the environmental chamber. The chamber, a 22-meter long by 3.66-meter wide room, houses a 19.5-meter long by 0.46-meter wide by 0.60-meter deep water channel. Down the length of the channel are 16 rotating paddle wheels which can be turned at speeds from 2 to 20 revolutions per minute. The purpose of these transport controllers is to regulate the level of turbulent mixing in the channel. Suspended over the channel is a radiant energy system which consists of narrow band red, green, and blue fluorescent lights and two sets of incandescent lights as illustrated in Figure 2. Since the lights radiate over narrow wavelength bands, the intensity of light in the blue, green, and red regions can be regulated independently. An air circulation system coupled with air chillers and heaters provide both air temperature and humidity control.

Water is presently supplied to the channel after passing through a deionizer, still, and heat exchanger. Since there is a 15-cubic meter storage capacity available, water from various field sites could also be tested. Table 1 summarizes the control range of environmental parameters varied in the AEcoS.

The analytical chemistry laboratory (Figure 1) houses most of the instrumentation used to monitor water quality in the water channel. Water samples from nine locations down the length of the channel are pumped through a completely automated system of valves which split samples into nine streams which in turn flow into nine auto-analyzers. The valving system is arranged such that as a water sample from one location is being analyzed, the flow from the other eight streams is diverted to collection bottles. The water collected in these bottles is used for biological analysis. Thus nine chemical parameters and a number of biological parameters can be recorded and studied simultaneously. The chemical tests which presently can be performed are shown in Table 2. Manual sampling techniques are used to measure total inorganic and organic carbon.

Dissolved gases are also monitored in the laboratory. The technique being adapted has been used in the medical field to measure blood gases in veins and arteries. Helium, a carrier gas, is passed through stainless steel tubing to a probe (a section of silastic tubing), submerged in the

[1] J. W. Falco and W. M. Sanders III are with the National Pollutants Fate Research Program, Southeast Environmental Research Laboratory, National Environmental Research Center-Corvallis, U.S. Environmental Protection Agency, Athens, Georgia. Mr. Falco's current address is U.S. Army Corps of Engineers, Waterways Experiment Station, Vicksburg, Mississippi.

water channel at a sampling location; from there the helium plus sample flows back to a gas chromatograph in the laboratory. As the helium passes through the silastic, it picks up oxygen, carbon dioxide, and nitrogen which is diffusing in from the water. These small amounts of diffused gases can be detected by the gas chromatograph and at the present time the method provides accurate estimates of oxygen and nitrogen concentrations in water. Because of the low level of carbon dioxide in water, carbon dioxide determinations are less precise and will require more development work. The obvious advantage of this type of probe over others is that with one probe three parameters are measured.

The control room (Figure 1) contains visual monitors and manual and mechanical controllers for all manipulatable environmental parameters. A CRT screen has also been mounted and connected to a PDP-8/e computer so that all environmental parameters measured can be monitored. A keyboard connected to the computer

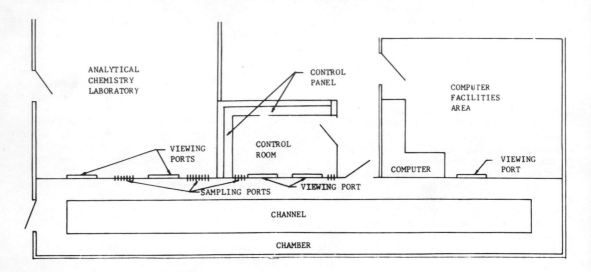

Figure 1. Floor plan–schematic of chamber facilities.

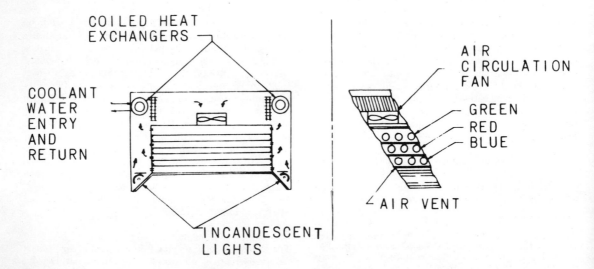

Figure 2. Schematic of ecosimulation chamber light source.

Table 1. Operational control specification.

Parameter	Range	Tolerance
Air Velocity	> 4 fps	
Air Temperature	0°C to 40°C	± 0.55°C
Relative Humidity	20% to 95%	± 2.0%
Influent Water Temperature	1°C to 40°C	± 0.5°C
Influent Water Flow Rate	200 gpd to 2000 gpd	± 2.0% of Flow

Table 2. Parameters to be measured in AEcoS experiments.

Parameter	Instrument
Total Phosphorus	Autoanalyzer
Ortho-Phosphorus	Autoanalyzer
Chemical Oxygen Demand	Autoanalyzer
Chloride	Autoanalyzer
Ammonia	Autoanalyzer
Nitrate	Autoanalyzer
Nitrite	Autoanalyzer
Kjeldahl Nitrogen	Autoanalyzer
Glucose	Autoanalyzer
Total Organic Carbon	Carbon Analyzer
Total Inorganic Carbon	Carbon Analyzer
pH	Probe
Temperature	Probe
Dissolved Oxygen	Probe
Dissolved Carbon Dioxide	Probe
Dissolved Nitrogen	Probe

is also located in the control room so that control functions for automatic control of environmental parameters can be altered by the operator. The control room also houses electronic components which are required to provide accurate easily analyzed transmission signals from various instruments and probes to computer storage and data handling systems.

The computer facility consists of a PDP-8/e computer and associated peripheral equipment. It has two major functions. First, in the normal mode of operation, it provides set points for all environmental parameters, including light intensity; and secondly, it serves as the data acquisition system for all environmental parameters measured.

The Physical Model and Preliminary Experiments

AEcoS is a process model in which processes are scaled rather than dimensions. Of the possible parameters that can be controlled perhaps the most important is level of turbulence. The level of turbulent mixing affects the rate of reaeration, alters residence time distribution in the channel, alters gradients of reactive materials along the length of the channel and affects the rate of nutrient uptake by microorganisms.

Since two parameters, rate of flow and transport controller rotation speed, can be manipulated within reason, the level of turbulence in the channel can be adjusted to match any two of the parameters listed above. The writers have chosen to simulate residence time distribution and reaeration rate in the first series of experiments for a number of reasons. From a matter of practicality there is a large amount of data on rates of reaeration and dispersion in published literature and government reports. Scaling the rate of dispersion and reaeration should ensure that concentration gradients of similar magnitudes occur in the simulation and in the natural ecosystem given equal average residence times and identical reaction rates. Scaling of reaeration rate should also normally match simulation and natural ecosystem mass transport limitations to nutrient uptake by microorganisms if surface area to volume ratios in both cases are the same (assuming there is a correspondence between bacterial growth and oxygen depletion).

Having chosen rate of dispersion and rate of reaeration as the two transport properties to be scaled, these two parameters must be measured quantitatively as functions of flow rate and transport controller speed, and once these functional relationships are established, these rates should be tabulated in dimensionless form for scaling purposes. In the case of dispersion, a series of dye tracer experiments has been completed which quantitatively specifies the dispersion coefficient D as a function of paddle wheel rotation speed and water flow rate. The experimental procedure followed is quite simple. After setting paddle wheel rotation speed and flow rate, a known amount of dye was injected at the center of the water channel. The concentration of dye was then monitored for various periods of time at three downstream locations. Two typical concentration profiles are shown in Figures 3 and 4. At high paddle wheel rotation speeds and low flow rates, high dispersion rates prevail. Consequently, the channel is well mixed and concentration profiles at different downstream locations are similar, as shown in Figure 3. At high flow rates and low paddle wheel rotation speeds the channel is poorly mixed. The dye moves down the channel essentially as a plug and concentration profiles at different locations do not match as shown in Figure 4.

If one mathematically treats the injection of dye as an impulse and carries out a mass balance on the spreading dye, the concentration profile should be predicted by the equation

$$c = \frac{M}{A\sqrt{4\pi Dt}} \exp\left\{\frac{-(x - vt)^2}{4Dt}\right\}$$

in which

c	=	concentration of dye
x	=	distance from the point of injection

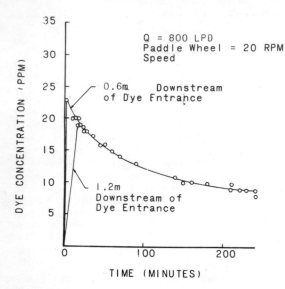

Figure 3. Dye penetration study.

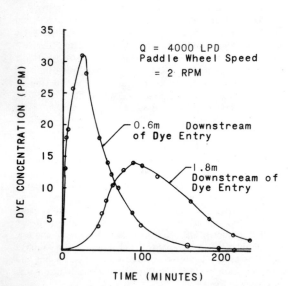

Figure 4. Dye penetration study.

M	=	amount of dye injected
t	=	time
D	=	dispersion coefficient
v	=	velocity

Since all of these parameters are known except D, a least squares fit of the data can be made to calculate a best fit for D. Figure 5 shows calculated dispersion coefficients as functions of flow rate and paddle wheel rotation speed. As these plots indicate, the rate of dispersion increases with increasing paddle wheel rotation speed and with increasing flow rate. Since the AEcoS is to be used as a scaled model of a flowing stream, it is advantageous to convert the dispersion coefficients shown in Figure 5 to a dimensionless measure of turbulent mixing. Such a dimensionless measure is the Peclet number (Kramer and Westerterp, 1973). This number is defined as

$$N_{Pé} = \frac{v \cdot l}{D}$$

in which

v	=	average bulk velocity
l	=	length of river reach

In Figure 6, the Peclet number is plotted as functions of flow rate and paddle wheel rotation speed.

Using similar principles, a series of experiments can be done to measure gas exchange rate at the air water interface as functions of temperature, flow rate, and paddle wheel rotation speed. In this case, influent channel water is sparged with nitrogen to remove oxygen. Then as the water moves downstream, oxygen is transferred into the water and nitrogen is transferred out of the water.

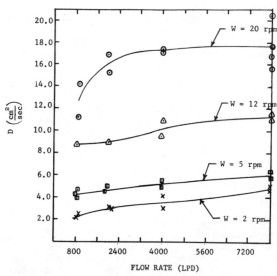

Figure 5. Variation in dispersion coefficients as a function of flow rate and transport controller speed.

162

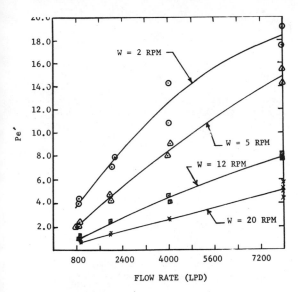

Figure 6. Variation in Peclet number as a function of flow rate and transport controller speed.

Carrying out a mass balance on transported gases yields

$$\left(\frac{C - C_{eq}}{C_o - C_{eq}}\right) = \exp\left\{\left(\frac{v - \sqrt{v^2 + 4DK}}{2D}\right)x\right\}$$

in which

C_o = inlet concentration

C_{eq} = concentration of dissolved gas that would be in equilibrium with air

x = downstream distance

K = gas exchange coefficient

Since all parameters except K are either known or measured, a least squares fit of the data to calculate the gas exchange coefficient can be made. Again since scaled dimensionless measures of gas exchange rates are desired, K should be expressed as

$$k = \frac{K\tau}{N_{Pé}}$$

in which

τ = average residence time

k = dimensionless exchange rate

Biological Rates—Spatial Scaling

The question arises as to whether one can spatially scale biological phenomena from a 19.55-meter long channel to a river reach which may be miles in length. The basis for spatial scaling is that phenomena occurring on the molecular scale occur over such small distances that movement of materials, chemical interactions, and other physical phenomena take place in a continuum rather than discretely. The assumption is that even though there are fewer molecules in the scaled down version of a model, there are still sufficient numbers of molecules which provide a basis for exercising the continuum approach.

In the case of microbiological phenomena, distances of microns rather than angstroms are involved. Even so, scaling up by 10^{12} the volumes of samples required for valid continuum samples, a 4,000 liter capacity channel would behave as a continuum. The one exception to this statement is the movement of motile organisms which cannot be scaled.

A second question arises concerning the possibility of completely different responses of the model and natural ecosystem to the same perturbation due to the limited number of species contained within the model. This is a weakness in both mathematical and physical models. Caution must be exercised in extrapolating data to natural ecosystems with similar environmental conditions and perturbations. In simulations care must be taken to insure major components (in terms of ecological niches) are represented.

Some Experimental Observations

To give some indication of the capability of the AEcoS, the results of a biological demonstration will be discussed. Procedures used in this demonstration were:

1. The channel was filled with distilled water to which sufficient amounts of materials were added to bring the concentration of required nutrients to values shown in Table 3.

2. Paddle wheel rotation speed was set at 2 revolutions per minute.

3. Air and water temperatures were set at 23°C.

4. Relative humidity was set at 50 percent.

5. Total light intensity was set at approximately 200 foot-candles. The lights were operated on a 24-hour cycle—12 hours on, 12 hours off.

Table 3. Concentration of nutrients in the AEcoS channel.

Chemical Constituent	Concentration
KNO_3	0.005M
K_2HPO_4	0.0005M
KH_2PO_4	0.0005M
$MgSO_4$	0.002M
$Ca(NO_3)_2$	0.00025M
Arnon A[5] Trace Element Solution	1 mℓ/ℓ

6. On the first day, six liters of a culture of *Chlorella pyrenoidosa* were inoculated into the upper end of the chamber.

The algae began to grow and spread downstream immediately, and after seven days, the entire channel contained a homogeneous green algal culture. As the culture became noticeably thicker, the light intensity was gradually increased to 4000 foot-candles. Maximum light intensity was set seven days after the original inoculation. At the end of the second week of growth, when it appeared that the maximum concentration of algae had been reached, the oxygen concentration was measured at 10 ppm and the carbon dioxide concentration was measured at less than 0.1 ppm during the light period. The fact that the system was carbon limited should not be surprising since all other nutrients were added in excess.

To provide more carbon dioxide for algal growth above and beyond the amount supplied by gas exchange at the air-water interface, six liters of a mixed bacterial culture were inoculated into the lower end of the channel. The bacteria used were originally cultured from water samples taken in Eighteen Mile Creek near Clemson, South Carolina. Glucose in quantities to bring the concentration up to 250 mg/ℓ was added at the same time the bacteria were inoculated into the system. The response of the ecosystem to the addition of bacteria was immediate. The algal population increased rapidly as indicated by a deep green color which spread up the entire length of the stream in less than seven days.

Peak algal populations were approximately 10^{10} cells/mℓ. Carbon dioxide concentrations remained less than 0.1 ppm and oxygen concentration remained near 10 ppm during light periods. After the population had stabilized, flow of nutrient was initiated in the channel. A flow of approximately 300 ℓ/day was maintained without washing out the cultures.

As a final test, an attempt was made to force the ecosystem to go anaerobic. To eliminate photosynthetic production of oxygen, the radiant energy system was turned off for 64 hours. At the end of this period, the dissolved oxygen concentration was measured and although the system was not anaerobic, oxygen concentration had dropped to 4.5 ppm.

Although this initial demonstration was not quantitative, it showed that a simple ecosystem could be established and maintained for an indefinite period, that the system could be manipulated to alter the chemical quality of the water, and that concentrations of algae could be obtained over ranges which are of environmental interest.

Mathematical Modeling

Our mathematical modeling efforts to date have been subordinate to the physical model construction; however, subroutines are required to analyze data from the physical model to provide experimental results in useful forms such as specific growth rates, maximum growth rates, mass transfer coefficients, etc. Such results can then be correlated with nutrient concentration, Peclet number, Reynolds number, and other pertinent variables. Assuming that the simulation is successful, these functional relationships can be applied to full-scale natural ecosystems.

Since the transport features of the AEcoS model are well defined, modeling the movement of materials through the channel is based on mass balances. Materials are partitioned into a bulk convective component and a dispersion component. Then according to a mass balance

rate of accumulation = net rate of convection + net rate of dispersion ($\pm$) source or sink terms

Since both source and sink terms are in general nonlinear, mass balances must be solved by finite difference techniques which give approximate solutions for specific initial and boundary conditions. As an example, Figure 7 illustrates the response of a flowing stream to a step change in concentration of a conservative substance. The technique used to solve the equation of continuity is outlined in a number of numerical methods texts (Salvadore and Bacon, 1971). Basically all spatial derivatives are approximated by central differences and the time derivative is approximated by forward difference. Integrations are ordered such that spatial integrations at a given time are accomplished and then a time step integration is carried out.

Mathematical models describing the uptake of nutrients and growth of microorganisms within the AEcoS remain to be developed. However, the intent is to apply major components of the ecosystem model being developed by Lassiter and Kearns (1973) for important chemical and biological processes occurring in aquatic ecosystems.

One other point of interest is the planned use of stochastic models to simulate environmental parameters for which field data have been collected. For example, solar energy data have been collected at the Southeast Environmental Research Laboratory at Athens, Georgia, between 1969 and 1972. The data, taken at weekly or biweekly intervals, include spectral distribution as well as total intensity measurements of solar radiation taken at frequent intervals throughout each test day. A typical spectral scan is shown in Figure 8. Obviously after three years of taking data at frequencies of 2 to 4 times a month, an enormous data set exists.

The radiant energy control system in the AEcoS has the capability of controlling independently the intensity of red, green, blue, and infrared lights. The spectral curves for those lights are shown in Figures 9, 10, 11, and 12. Figure 12 shows an unbalanced output of the four lamp types; however, it does show the infrared spectral output. To obtain a control function which gives the best fit of the simulated solar energy spectrum to a natural solar energy spectrum both in terms of distribution and total intensity, a stochastic model of the spectral distribution and intensity of solar radiation as function of day of the year and time of the day has been undertaken.

The first step has been to empirically model the distribution of sunlight as a function of wavelength. After some trial and error, it was determined that a piecewise fit of the spectral curve which was separated into three parts provided an adequate fit. The form of the equation used for each segment of the fit is

$$\frac{dI}{d\lambda} = \frac{C_1 S_1 \exp\left\{-5\left(\frac{\lambda - u_1}{S_1}\right)^2\right\}}{\sqrt{6.28}}$$

in which

I = intensity of solar radiation
λ = wavelength of radiation

C_1 = constant
S_1 = constant
u_1 = constant

A typical fit to a solar spectral curve is shown in Figure 13. Since in general there are no more than three large peaks in the solar spectrum, it is believed that a three-segment fit will provide adequate simulation detail for any given time.

Variations of spectra with time of the day and day of the year remain to be modeled. When this task is accomplished patterns of cloud cover development and concomitant reductions in light intensity will be modeled stochastically.

Conclusions

This paper gives a brief description of the Aquatic Ecosystem Simulator and an outline of the purpose and capabilities of this physical model. The experimental program for which this facility is designed is just beginning. In addition to the physical calibrations described, microbial growth experiments with bacterial and mixed bacterial and algal populations will be performed to

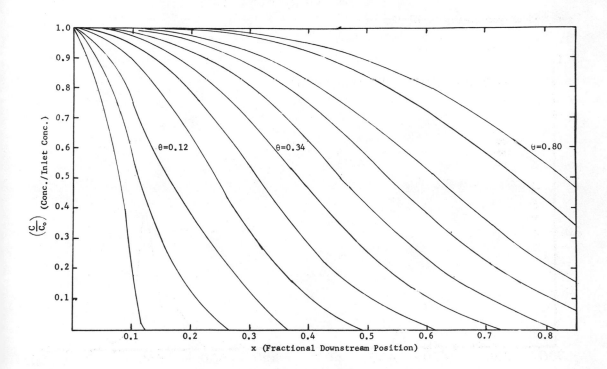

Figure 7. Response of a flowing stream to a step change in concentration (θ = time).

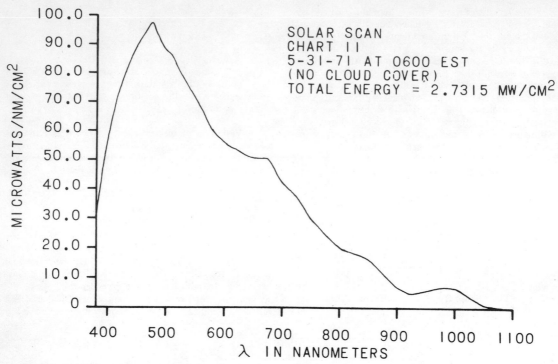

Figure 8. Solar spectral energy distribution curve.

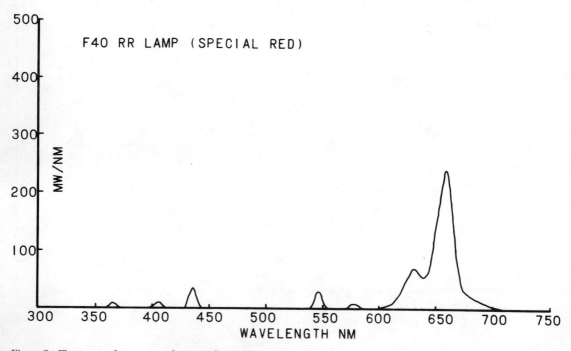

Figure 9. Fluorescent lamp spectral energy distribution curve.

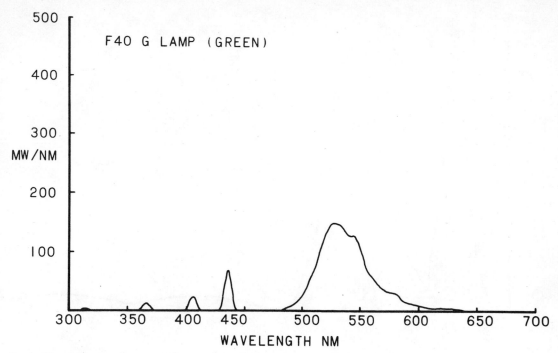

Figure 10. Fluorescent lamp spectral energy distribution curve.

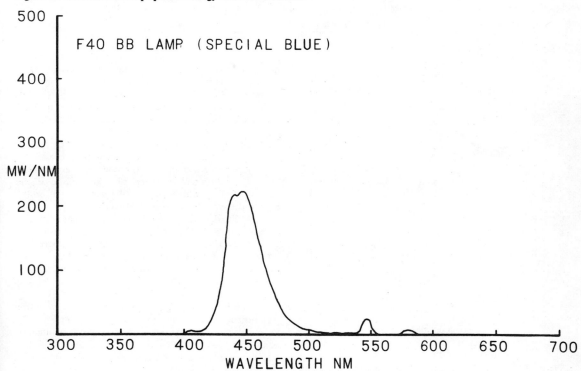

Figure 11. Fluorescent lamp spectral distribution curve.

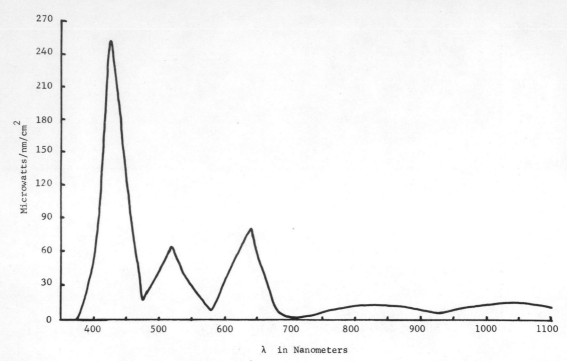

Figure 12. **Spectral scan from the AEcoS radiant energy system.**

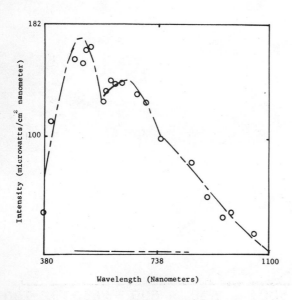

Figure 13. **Fit of solar spectrum piecewise to three normal distribution functions.**

measure growth rates of these organisms as functions of turbulent mixing rates, water temperature, nutrient concentration, and other pertinent environmental parameters. Ideally, parameter variations would match observed conditions from a field study on a natural body of water.

Conducted in this manner, the results of such an experimental program would accomplish three goals:

1. Provide a data set from controlled experiments to test validity and accuracy of deterministic mathematical models which predict the response of simple aquatic ecosystems to environmental parameters.

2. Provide the operational experience needed to account for various "process distortions due to scale."

3. Provide estimates of rate coefficients needed to predict short-term and long-term ecosystem responses to various environmental perturbations and increases in nutrient loads.

The last two program objectives are the incentives to construct and operate this rather costly facility. With increased public awareness of the need to maintain an acceptable water quality on one hand and a need to maintain a viable and competitive national economy on the other hand, it is imperative that methods to accurately describe water quality dynamics be developed. This knowledge should provide a sound basis for adequate legislation to balance the goals of a clean environment and sound economy.

References

Chuck, F.J. 1970. Mass transfer to tubes in the cardiovascular system. M.S. Thesis, University of Florida.

Kramer, H., and K.R. Westerterp. 1963. Elements of chemical reactor design and operation. Academic Press, Inc., New York, pp. 74-77.

Lassiter, R.R., and D.K. Kearns. 1973. Phytoplankton population changes and nutrient fluctuations in a simple aquatic ecosystem model. Symposium on Modeling of the Eutrophication Process, Logan, Utah.

Phinney, H.K., and C.D. McIntire. 1965. Effect of temperature on metabolism of periphyton communities developed in laboratory streams. Limnol. Oceanog., 10:341.

Salvadore, M.G., and M.L. Bacon. 1971. Numerical methods in engineering. Prentice-Hall, Inc., Englewood Cliffs, New Jersey, Second Edition, pp. 80-87.

Sanders, W.M., and J.W. Falco. 1973. Ecosystem simulation for water pollution research. Advances in Water Pollution Research, Pergamon Press, New York, pp. 243-253.

ACTIVITY ANALYSIS AND THE MANAGEMENT OF RESOURCES:
A MODEL FOR CONTROL OF EUTROPHICATION[1]

D. B. Porcella, A. B. Bishop, and W. J. Grenney[2]

Introduction

Resource management and control

Control of nutrient input to surface waters as a means of reversing eutrophication (e.g. Edmundson, 1972; Laurent et al., 1970; Liebman, 1970; Mackenthun et al., 1960; Oglesby, 1969; Porcella et al., 1972) is one example of resource management. However, for such management to have practical long-term benefits to society a rather complete picture of the consequences of control actions and the effects of choosing particular alternatives must be obtained. Environmental problems involving resource management usually have to exert significant effects before there is sufficient attention to bring about corrective action. The typical chain of events in corrective action is illustrated in Figure 1; however, in many cases naive solutions which may be counterproductive are developed irrespective of possible long-term social and economic "costs." The development of control strategies for eliminating the problem should be based on a rational development of relationships between the problem, the factors contributing to the problem, and the "costs" (social, economic, technological) associated with the solution to the problem.

Environmental management of particular resources implies that the system in which a resource is distributed is well understood and the mechanisms of distribution are well known. The phosphorus resource is relatively well described; phosphorus control is of interest because of its important role in the development of eutrophic conditions in lakes. Because of this role many suggestions have been made towards the control of eutrophication by elimination of some of the phosphorus input to surface waters. Proposed approaches have included phosphorus removal from detergent formulations, removal of phosphorus from domestic wastes, and the application of the concept of zero discharge from point sources. These manipulations of phosphorus input are considered rather simplistic solutions of complex problems. The immediate questions raised are first, whether these kinds of manipulations will produce improvement in the conditions, and second, whether in fact these solutions may be counterproductive in terms of the total economy of a region or the nation.

In response to these questions, this study was aimed at achieving a relatively complete understanding of phosphorus cycling in the social-technical-ecological system to be used as a basis for developing a strategy for phosphorus control which would decrease eutrophication effects at a minimum cost. A cost-effectiveness point of view, i.e., maximizing the effectiveness of control of phosphorus input to surface waters at least cost, was adopted because of the difficulty in quantitating the economic values of the various benefits of minimizing eutrophication. Evaluation of whether benefits of phosphorus restriction in natural waters justify those costs, then, is yet a value judgment which must be made in the political decision-making process.

The objectives of this study were, first, to describe the system in which phosphorus is utilized, second, to determine the possible alternatives for controlling phosphorus input to surface waters, and third, to find the least-cost strategy for controlling that phosphorus input. These objectives were to be applicable on a basin-wide or regional basis.

Phosphorus and eutrophication

Because the possibility exists for control of eutrophication by minimizing phosphorus input to the surface waters, this study has been directed toward the management of phosphorus. Before beginning an analysis of phosphorus using activities and the costs of its control, a relationship between phosphorus input and the problem—eutrophication—is needed. Such a relationship, based on Sawyer's initial work (1947), had previously been developed by Vollenweider (1968). Schindler et al. (1971) have calculated a log functional relationship for "admissible (A)" and "dangerous (D)" nitrogen (N) and phosphorus (P) levels between nutrient loading ($g/m^2 \cdot$ yr) and

[1]Work described herein was supported by the Office of Research and Monitoring, U.S. Environmental Protection Agency, under Contract No. 68-01-0728, Washington, D.C. 20460.

[2]D.B. Porcella, A.B. Bishop, and W.J. Grenney are with the Utah Water Research Laboratory, Utah State University, Logan, Utah.

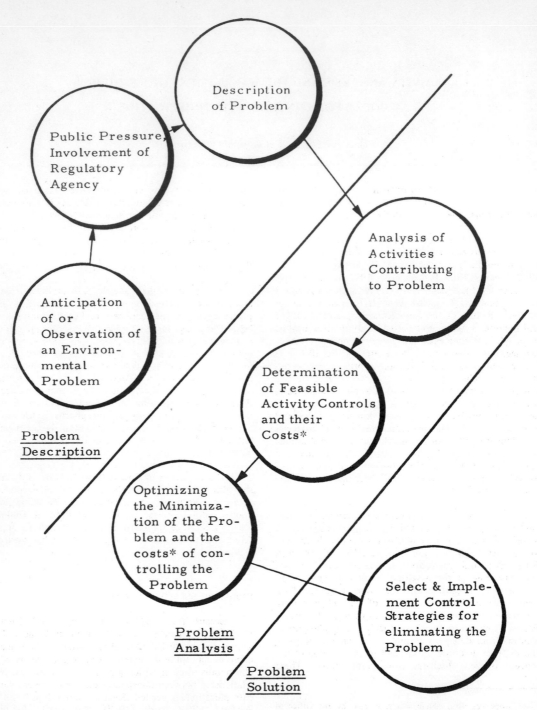

Within the figure:

Description of Problem

Public Pressure, Involvement of Regulatory Agency

Anticipation of or Observation of an Environmental Problem

Analysis of Activities Contributing to Problem

Determination of Feasible Activity Controls and their Costs*

Optimizing the Minimization of the Problem and the costs* of controlling the Problem

Select & Implement Control Strategies for eliminating the Problem

Problem Description

Problem Analysis

Problem Solution

*"Costs" include technological, implementation, and social costs.

Figure 1. General scheme for the stepwise solution to environmental problems.

the mean depth ($\overline{Z}$) of the receiving water (based on Vollenweider, 1968):

$$\left.\begin{array}{l} \log_{10} P_A = 0.60 \log_{10} \overline{Z} + 1.40 \\ \log_{10} P_D = 0.60 \log_{10} \overline{Z} + 1.70 \end{array}\right\} \quad \cdots \quad (1)$$

$$\left.\begin{array}{l} \log_{10} N_A = 0.60 \log_{10} \overline{Z} + 2.57 \\ \log_{10} N_D = 0.60 \log_{10} \overline{Z} + 2.87 \end{array}\right\} \quad \cdots \quad (2)$$

In these formulations oligotrophic lakes were found to occur at loadings below admissible levels while eutrophic lakes occurred at loadings above dangerous levels; presumably mesotrophic lakes were between the admissible and dangerous loading levels.

Both Vollenweider (1968) and Schindler et al. (1971) have suggested considerable caution in the use of these relationships because of the lack of confirmatory studies and the great error in using such a simple formulation for the description of such a complex and highly variable system. With this warning in mind, further extension of Vollenweider's loading estimate will be made to obtain provisional guidelines of the relation between influent phosphorus and eutrophication.

For lakes having different fixed mean depths ($\overline{Z}$), Equation 1 gives an estimate of the two points, admissible and dangerous, on the eutrophication-phosphorus relationship. Thus, an annual phosphorus loading rate can be calculated from a phosphorus input model and related to the mean depth of a particular lake using the relationship in Figure 2. Relative eutrophication is measured as an index number where the smaller the number, the less eutrophic the water body. The effect on phosphorus input by a particular management scheme is related to levels of eutrophication and the results evaluated in terms of the cost of the management scheme. This index number is not meant to imply a linear or any other functional relationship; for example, one might expect a first order-zero order relationship.

These eutrophication estimates can be recast in terms of chlorophyll a concentrations and phosphorus loading to allow an extremely crude lake measurement of admissible and dangerous levels of eutrophication. The relationships can be determined by substituting in turn 1) the relationship between spring phosphorus concentrations and annual phosphorus loading rates (Vollenweider, 1968); 2) the relationship between chlorophyll a and the winter-spring phosphorus concentration (Edmundson, 1972; Megard, 1972); and 3) relationships for Secchi depth and chlorophyll a (Edmundson, 1972). The major assumption in making this substitution is that Vollenweider's spring phosphorus concentration is the same as the winter-spring concentrations described by Edmundson and Megard. As shown in Figure 3, the relation between phosphorus loading rates and chlorophyll a and Secchi

depth allows definition of "lake biota" levels corresponding to admissible and dangerous levels of eutrophication. Both calculated levels allow quite high algal populations and although of some use as a relative measure to indicate in quantitative terms a crude definition of algal population levels corresponding to a eutrophic lake, they are unrealistic and probably not acceptable; however, they can be compared to other lake systems varying in their degree of eutrophy.

Activity Analysis

Basic assumption of this approach

This approach, and any other similar resource management analysis, is based on the assumption that there is a quantifiable relationship between the environmental effects of the resource and the uses made of that resource. For phosphorus it was assumed that control of its input to natural waters will cause a reduction of eutrophication as shown in Figure 2. Consequently, a comprehensive understanding of phosphorus uses—an activity analysis—is required before logical management schemes can be attempted.

Summary of phosphorus sources

Natural sources. Ultimately all phosphorus comes from minerals. These minerals are weathered by natural processes or utilized by man and other organisms causing the release of phosphorus for recycling through biological communities or for transport to the oceans. After entering a lake or ultimately the ocean, the processes of biological deposition, physical deposition, and chemical precipitation transfer phosphorus to the bottom sediments where it is stored until geologic time restores it to the land surface for further utilization.

Although there are many natural processes which serve as sources of phosphorus to surface water, human activities, at least in the USA, probably account for the majority of phosphorus mass inputs. Natural inputs consist largely of solution effects as water passes over and weathers geological formations. Some phosphorus becomes cycled through biological materials prior to entering water systems as inorganic or organic phosphorus compounds, but ultimately, these biologically derived compounds come from phosphorus minerals also.

Cultural use of phosphorus. Cultural processes cause an increase in phosphorus input to surface waters in comparison to the natural inputs normally expected for a particular system. Except for direct food plant uptake of natural soil phosphorus, essentially all use of phosphorus by human society depends on the phosphorus mining industry. In 1968, the estimated distribution of phosphorus obtained from mining activities was primarily to agriculture (about 80 percent used in fertilizers and commercial animal feeds), with a considerably smaller

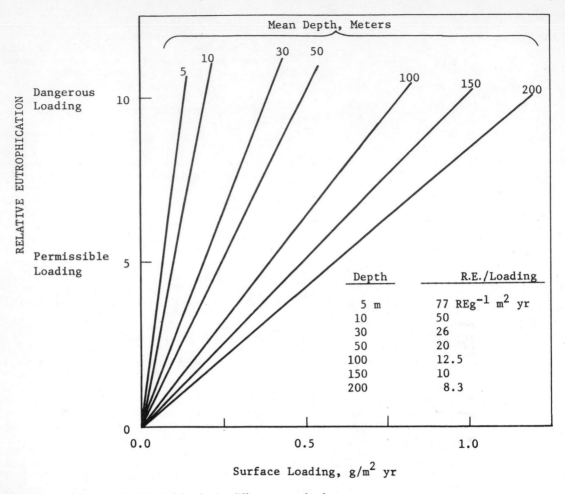

Figure 2. Relative eutrophication in lakes having different mean depths.

amount (7 percent) used as phosphate builders in the detergent industry and the remaining 13 percent used for various industrial purposes and as product additives (see Table 1).

Future changes in the usages of phosphorus, particularly fertilizer and animal feeds, detergent, and industrial uses, will depend primarily on population changes, changes in agricultural practices (chiefly due to urbanization of agricultural lands and international trade policies), and possible restrictions on detergent phosphorus uses and on industrial uses. Technological developments will probably have minimal effect on the use of phosphorus products.

Activity analysis of phosphorus

Using an activity analysis, a system was developed which could be analyzed from the point of view of

phosphorus control; this included a mass flow model for the major phosphorus-using activities and natural sources which could be applied to a particular region or surface water basin (Porcella et al., 1973). The model, utilizing input information that is generally available, calculates phosphorus output according to different activities. This basic approach is summarized in Figure 4. Some of the activities are further affected by various types of treatments or distributions depending on whether wastes are directly discharged or enter treatment plants. The total of all the activities is calculated as well as the total after treatment, or that portion of the phosphorus which actually enters the surface water. Finally, the total of the fraction entering the surface water which is actually available for algal growth is calculated. Calculation of eutrophication effects due to phosphorus addition to surface waters is based on two loading rates (total after treatment, and total available for algal growth) because of the uncertainty of the defined relationship between

Relative Eutrophication	Biomass Parameter	
	Chlorophyll a, µg/l	Secchi Depth, m
Permissible	26	1.02
Dangerous	52	0.85

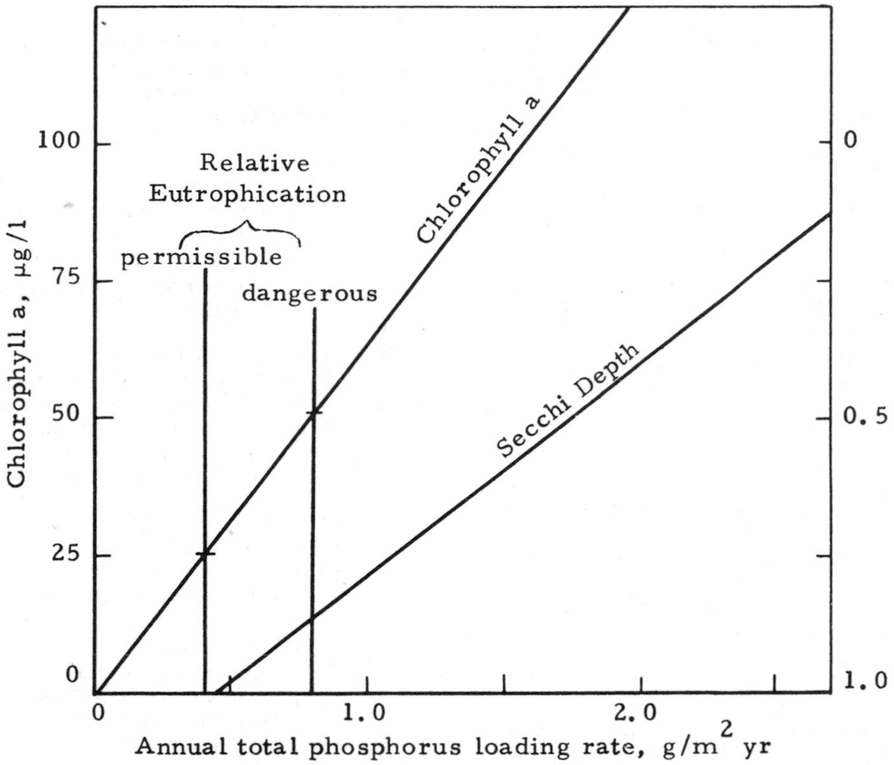

Figure 3. An example of how a biomass (algal bloom) parameter can be related to phosphorus loading rate. Assumed mean depth of lake is 100 meters for estimation of relative eutrophication.

nutrient loading rate and the availability of the different forms of phosphorus for algal growth and eutrophication (Vollenweider, 1968). Note that treatment refers to the use of tertiary processes added to existing plants; these are coagulation, ion exchange, and reverse osmosis.

The detailed activity analysis in Figure 5 was developed using the information contained in Table 1 and considering other probable phosphorus inputs to surface waters. These activities have been grouped into 1) non-basin activities; 2) agriculture; 3) urban and rural watersheds; 4) domestic; 5) industrial; 6) mining; and 7) animal production. The pathways in the figure indicate that most of the activities involved with phosphorus entering surface waters are associated with human activities. The natural sources are quite variable and are dependent on many factors as will become evident in the description below. The cultural uses tend to be more constant but the inputs to surface waters vary with discharge regulations and geography for the diffuse sources.

Phosphorus using activities and methods of control

The use of phosphorus in various industrial, agricultural, and domestic activities plus natural inputs of phosphorus all contribute to surface water phosphorus content. For each of the major defined activities (Figure 5), appropriate and reasonable control points can be devised at the supply and demand side of the activity as well as for the technological and treatment side of the activity. The flow chart shown in Figure 5 has been combined together with information in Table 2 to

Table 1. Phosphorus consumption in the USA.

Uses	Amount Used, 1000 Kg/yr[a] as P (percent of total)	
	Year: 1958[b]	1968[c] (estimates)
Fertilizers	896 (69.7)	2406 (76.3)
Animal Feed	107 (8.3)	94 (3)
Detergents	171 (13.3)	227 (7.2)
Metal Finishing	15 (1.1)	94 (3)
Other Uses	97 (7.6)	331 (10.5)
Water Softening	35 (2.7)	35 (1.1)
Food and Pharmaceutical	38 (3)	58 (1.8)
Gasoline Additives Plasticizers Pesticides	8 (0.6)	29 (0.9)
Miscellaneous	16 (1.3)	210 (6.7)
TOTAL	1286 (100)	3153 (100)

[a]To obtain 1000 short tons of P/yr, multiply by 1.1013.
[b]Estimated from Logue (1959).
[c]Estimated from Lewis (1970).

complete the activity analysis, i.e., to identify and quantify as precisely as present literature allows all of the natural and cultural activities which tend to produce phosphorus inputs to surface waters in conjunction with possible control measures. By applying costs to each control, a cost-effectiveness relationship can be defined.

The analysis of the phosphorus inputs to a particular receiving water was based on particular subroutines and/or numbers obtained from the literature for those particular inputs. These were based on average values imputed to various activities. For example, 1) domestic wastes phosphorus loadings were placed on a per capita basis; 2) industrial wastes were based on particular industrial uses; 3) natural inputs, fertilizer runoff, pesticides and mining wastes were related to activities placed on an area basis; and 4) animal wastes solid wastes were on a per capita basis. Thus, for a given subbasin the input of information based on per capita and area estimates of the different activites would provide an input in terms of mass of material per unit time. This input was coupled to the receiving water eutrophication model of Vollenweider's shown in Figure 2.

That portion of the phosphorus input which could undergo the various waste treatment schemes currently feasible were subjected to a variety of tertiary treatments and the appropriate designated removal resulted in a particular direct cost. These costs and effects of treatment on phosphorus input and hence eutrophication were also determined in the computer program (discussion of this aspect is in Porcella et al., 1973) for the mass flow model.

Table 2. Summary listing of control tactics.

A. Supply and demand (applies to consumer habits and producer activities)
   1. Subsidies (non-phosphorus products)
   2. Tax breaks and credits
   3. Price controls
   4. Excise taxes or other taxes
   5. Advertising and education
   6. Non-monetary recognition
   7. Content labeling
   8. Moral suasion
   9. Boycotts

B. Resource control, mining and manufacturing
   1. Requirements for recycling
   2. Phosphate mining restrictions (rationing)
   3. Manufacturing/production restrictions
   4. Emission controls

C. Management of phosphorus uses
   1. Resource and product substitution
   2. Technology improvements in processes or uses
   3. Monitor requirements with enforcement of application rates (e.g., fertilizer)
   4. Recycling and reclamation

D. Management of phosphorus discharges
   1. Pollution standards
   2. Land management practices
      a. Reduction of cultivated acreage
      b. Increased fertilizer use
      c. Technical management
      d. Irrigation practices
      e. Green belts and buffer zones
      f. Solid waste recycling
   3. Land use controls
      a. Zoning
      b. Licensing
      c. Leasing
      d. Codes and subdivision regulations
      e. Permits
   4. Solid waste management
      a. Disposal regulation
      b. Fees
   5. Effluent charges
   6. Bans
   7. Fines

E. Judicial controls
   1. Judicial review
   2. Class action
   3. Common law remedies (nuisance trespass, negligence)

F. Wastewater treatment—for phosphorus removal

G. Lake modification

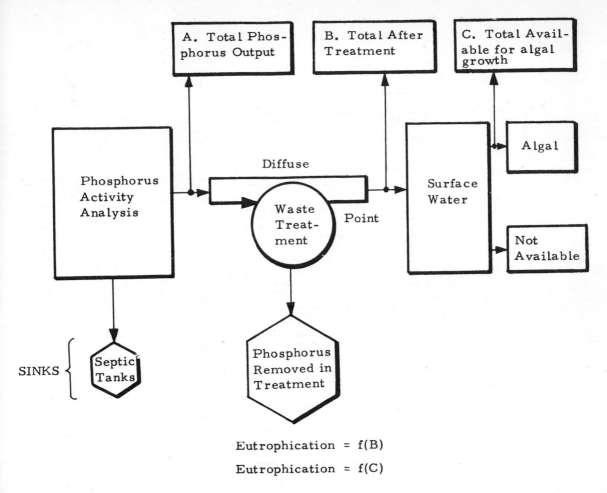

$$\text{Eutrophication} = f(B)$$

$$\text{Eutrophication} = f(C)$$

Figure 4. Primary calculated outputs from the phosphorus mass flow model.

Cost-effectiveness Analysis of Phosphorus Management Strategies

The outputs from the phosphorus mass-flow and eutrophication model, applied to water resource basins, provide the basic information needed for analyzing the cost-effectiveness of various strategies for phosphate management. A management strategy consists of a co-ordinated set of tactics (from Table 2) implemented to control phosphorus flows from the various activities (agriculture, urban and rural watersheds, domestic and industrial wastewaters, mining, and animal production) represented in the mass flow model (Figure 5). The effects of different strategies are examined by manipulating the model inputs to simulate the application of the specified controls to the phosphorus activities, then determining the change in eutrophication levels and the treatment costs associated with achieving those levels.

General framework for strategy evaluation

A comprehensive analysis of strategies for phosphorus management in surface waters would, in general, involve consideration of both the benefits and costs associated with achieving given levels of phosphorus reduction. The diagram of Figure 6 gives a general description of where costs and benefits arise in controlling phosphorus flow from a given activity, and shows how cost/benefit and cost/effectiveness evaluations are derived. Costs are associated with controls applied at various points in the system to reduce or eliminate amounts of phosphorus mobilized by an activity or amounts flowing into surface waters. Benefits are mainly realized in terms of economic, social, and environmental returns from water use, and the damage costs from eutrophication that are avoided. A benefit/cost comparison then tests the question of whether the cost is worth the benefits

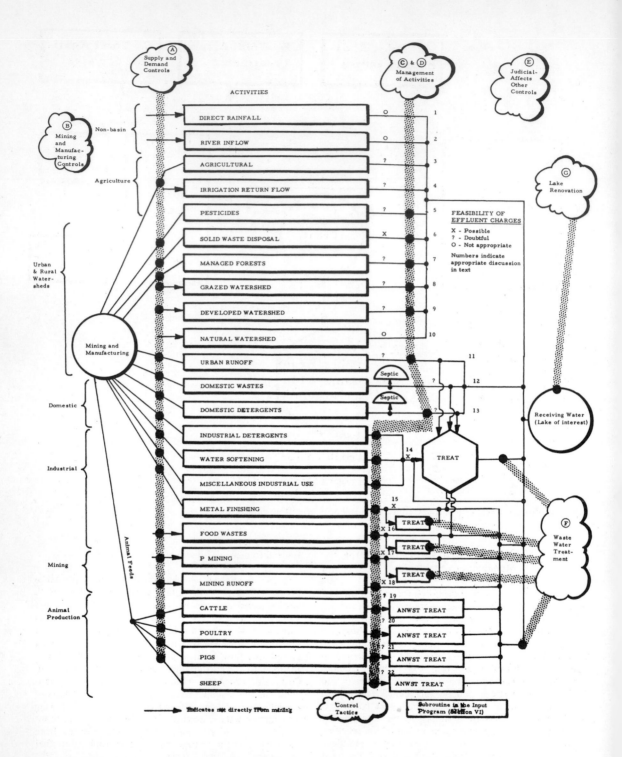

Figure 5. Control points superimposed on the phosphorus activity analysis showing the major application points for pertinent control tactics. (See Table 4-6.)

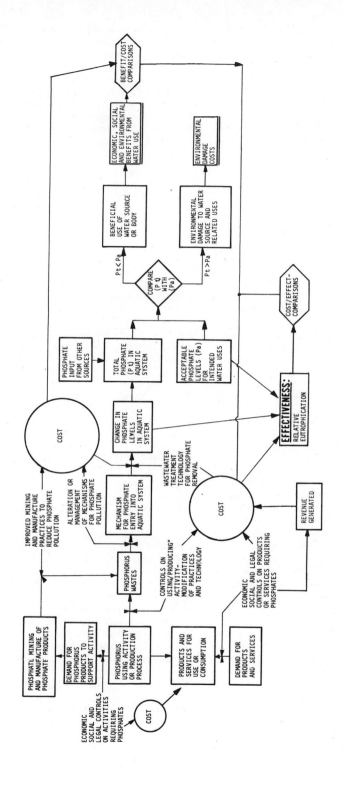

Figure 6. Cost-benefit and cost-effectiveness analysis related to phosphorus mass flow.

179

received. Unfortunately, the benefits from phosphate reduction are difficult to assess and at present no adequate data are available for doing so. However, cost-effectiveness analysis provides a rational basis for management decisions by comparing the cost of various alternatives with their effectiveness in achieving a given objective.

As Figure 6 shows, the objective for effectiveness was chosen to be measured in terms of changes in relative eutrophication (the primary purpose for phosphorus management) rather than actual reductions of phosphate flows into a receiving water.

The approach to cost-effectiveness analysis

In analyzing the costs to be used in evaluating cost-effective strategies, only "real" costs, i.e. those which divert resources from other productive uses in providing goods and services, are considered. In general, four kinds of costs are identified as real costs:

(1) *Production losses*—One way to reduce the flow of phosphorus into aquatic systems is to diminish production from phosphorus generating activities. Actions which diminish production give rise to real costs from production losses.

(2) *Production cost increases*—Production costs, including investment (or capital) costs and operating costs, are incurred directly by the producing firm. Treatment plants attached to an individual firm or to a group of firms or individuals are example actions which increase production costs.

(3) *Government program costs*—Costs of direct actions by government are, in many ways, comparable to the production costs increases noted above. Government program costs are appropriate for overcoming problems in which the sources are difficult to identify, or for other reasons the external effects cannot be internalized to the producer. Since the cost of these government programs represents an alternative to production of other private or public goods and services, they are a real cost as distinguished from government transfer payments.

(4) *Administrative costs*—For most types of controls, certain administrative costs for monitoring and supervising compliance will be needed, including personnel costs and instrumentation expenses. As with program costs these are real costs since other goods and/or services could be obtained if these expenditures were not made.

While all the elements of these costs appropriately belong in the cost-effectiveness analysis, to completely determine them would require a detailed cost analysis for each control measure based on specifications for its implementation. Furthermore, costs will also depend on the particularities of the basin area being considered. Because of these difficulties, the cost-effectiveness of various management strategies is analyzed in terms of cost-savings which can be accrued by avoiding the need to apply treatment in order to achieve a specified level of effectiveness of relative eutrophication. The following simplified example serves to illustrate the cost-effectiveness analysis approach.

In Figure 7 the relative eutrophication level of a lake is measured on the ordinate and the treatment cost associated with strategies for achieving those levels on the abscissa. Two cost-effectiveness curves, similar to those derived from the P-mass flow model, previously described, are plotted. Strategy A relies solely on treatment processes for all phosphorus removal; strategy B applies a set of management controls (tactics) along with treatment. The curves show that the effect of the management controls is to reduce the eutrophication level to R. To achieve the same level through the use of treatment technologies would have a cost of S dollars per year. The question is whether or not the management strategy should be implemented. The answer to this is stated in the form of the following decision rule: If the "real" cost of strategy implementation is less than S, then it pays to adopt the management controls. To further illustrate this, if the desired effectiveness in eutrophication level were Q, then the cost of achieving this through strategy A-Treatment only is "U" dollars per year. The cost-savings of implementing the management controls in achieving the same level of relative eutrophication, Q, is the amount T dollars per year. In this case, the management strategy should be implemented so long as the real costs are less than T dollars per year.

A case example: The Lake Erie basin

The following case example for the Lake Erie basin illustrates the model output and analysis approach for a real system. The phosphorus mass flow model and treatment subroutine are used to generate a set of cost-effectiveness curves for comparing alternative management strategies. If a desired achievement level for relative eutrophication which maintains a water body in a non-eutrophic state is given and decision-maker's estimate of the real costs of controls for his particular situation is given, a cost-effective strategy can be selected. Similarly, if the total program budget is fixed, then a strategy which maximizes the level of effectiveness given the budget constraint can be determined.

Targeting P-activities for application of control strategies

To identify those activities in the basin that are responsible for the major inputs of phosphorus to Lake

Erie, model runs were made for the presently existing conditions in the basin. Although the mass flow program was applied to Lake Erie without any special adjustments for curve fitting, reasonable agreement with the actual situations was observed. The calculated phosphorus concentrations of about 42 μ g P/l for Lake Erie compare fairly closely with estimates of 11-90 μ g P/l in filtered water from the lake (Lange, 1971). Thus, application of activity analysis and cost-effectiveness to the lake appears reasonable. The baseline condition for the basin with a given set of input data is shown in the bar graph of Figure 8. The chart shows the phosphate output produced by each activity as a percentage of the total output generated in the basin and readily identifies those activities that contribute the majority of the P-loadings to the lake. For a number of activities in the basin the output level is so slight as to be negligible, so these can be ignored insofar as the application of control strategies. On the other hand, major input sources for Lake Erie, such as domestic wastes and domestic detergents are prime targets for application of strong and effective management strategies to control eutrophication.

Application of comprehensive management strategies

After identifying the major input activities, different control strategies were applied in order to assess the results in terms of the relative eutrophication of Lake Erie (Figure 9). The particular strategies analyzed were: 1) to apply treatment processes to all phosphorus inputs from sources which can be treated. Zero removal establishes the base case from which cost savings are calculated. Percentage removal values are shown where this is applied; 2) to ban use of high phosphorus detergent, both in the domestic (DomDet) and industrial use (IndDet) patterns; 3) to apply land use controls and practices (AgMgmt), such as green belts and retiring of high-slope lands (greater than one percent), in order to reduce erosion rates and eliminate soluble surface phosphourus runoff; 4) to transfer all animal waste (ANWST-land) to land disposal and allow only 10 Kg/ha. yr as the maximum level of phosphorus fertilization; 5) to require all industrial waste to go through the sewage treatment plant; and thus, direct discharge to wastes to streams was prohibited ("zero

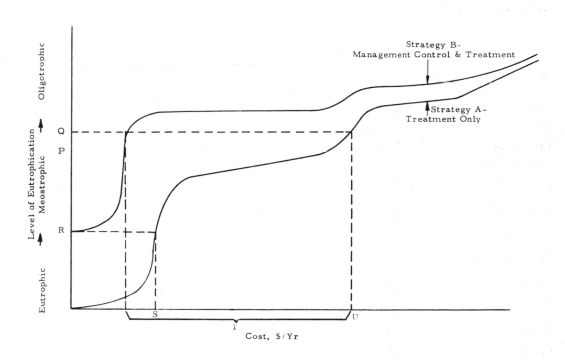

Figure 7. Example of cost-effectiveness curves.

181

discharge"); 6) to eliminate certain kinds of runoff conditions, and to minimize phosphorus use in road deicing compounds and in home garden use; these were assumed to halve the amount of phosphorus coming from urban runoff (zero minor); 7) to reduce all minor inputs and water softening (water soft) to zero.

Using these strategies alone and in combination, levels of wastewater treatment were then applied. The effect of these strategy combinations on relative eutrophication in Lake Erie is shown in Figure 9. The upper dotted line shows the present eutrophication level. A bar for each strategy then shows the resultant change in

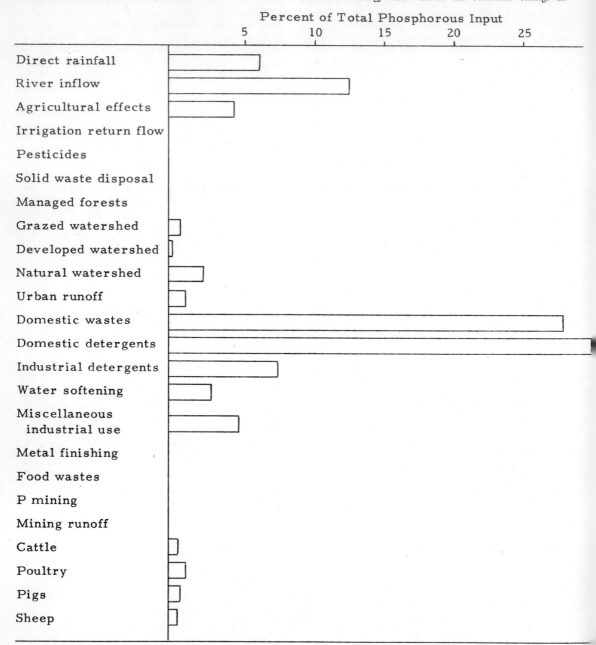

Figure 8. Lake Erie—relative contributions from phosphorus activities prior to application of controls.

eutrophication level by applying the specific management controls. The open space in the bar and the cross-hatched portion shows relative eutrophication due to total phosphorus loading and available phosphorus loading.

Cost-effectiveness analysis of strategy implementation

The cost-effectiveness analysis of implementing the set of control strategies screened out as being potentially most effective is developed along the lines suggested in the earlier generalized example (Figure 7). Data for the cost-effectiveness curves are generated by a series of model runs which simulate the effect of the management control together with wastewater treatments at specified percentages of phosphorus removal. For each level of removal, the relative eutrophication and treatment cost are computed. Costs are in dollars per year. Relative eutrophication and costs determine a point on the cost-effectiveness curve for that management strategy.

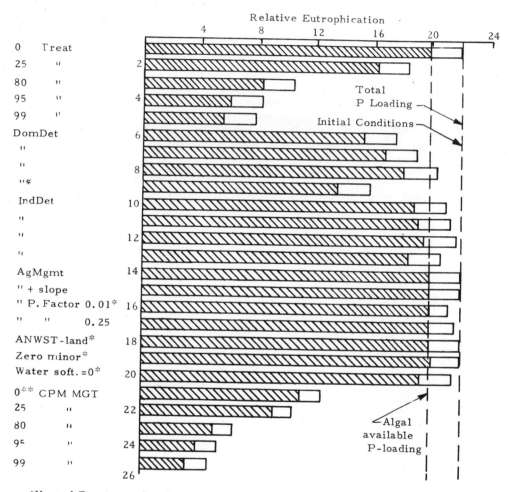

*Varied Treatment levels were later applied with all these strategies to form CPM MGT.
* Sewered all IND wastes. Urban Runoff Conc = 0.5 orig.; CMP MGT.

Figure 9. Lake Erie—effects of controls on relative eutrophication (see text for description).

183

The collection of points plotted for the percentage range for treatment removal generates the entire cost-effectiveness curve for the strategy. The cost-effectiveness curves for Lake Erie are presented in Figure 10 for the final set of management controls.

A general conclusion which can be drawn from the curves is that increased effectiveness from treatment can be obtained for lower treatment cost by applying management controls to sources in order to reduce phosphorus loadings. This, of course, was expected and confirms the validity of using the concept of cost-savings, or avoidance of treatment costs, as the basis of a decision rule for determining whether a management strategy should be implemented. To restate the rule: If the real costs of management controls are less than the treatment costs

saved (or avoided) then the strategy should be implemented.

Another general observation is that each cost-effectiveness curve reflects segments of cost functions from three different treatment processes, namely coagulation, ion exchange, and reverse osmosis. The treatment subroutine selects the minimum-cost treatment process based on constraints which specify acceptable influent concentration and the effluent concentrations which are obtainable from the process. When a higher-level treatment process is selected, such as ion exchange or reverse osmosis, there is a discontinuity in the cost-effectiveness curve (shown by the split line) which reflects the substantially higher cost function of the higher level process.

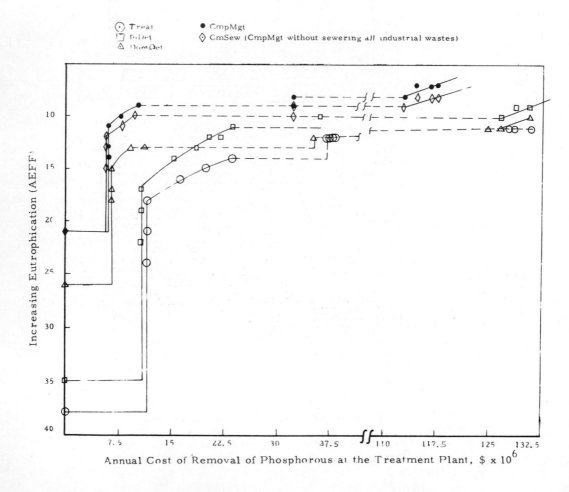

Figure 10. Lake Erie (20 m mean depth)—cost-effectiveness of various treatment levels in relation to eutrophication based on available phosphorus loading.

Lake Erie is a case in which a combination of management controls and treatment processes can, in fact, reduce the relative eutrophication levels to within an acceptable range of 5 - 10 for a lake.

In applying cost-effectiveness analysis, the selection of a strategy should be approached from the standpoint of the desired level of effectiveness and of the constraints on available budget. For example, consider a policy objective to achieve an effectiveness level of less than 10 to ensure non-eutrophic lake conditions. The cost-effectiveness curves of Figure 10 indicate that the management options can produce significant cost savings at this level. The curves show that a eutrophication number of around 10 is attainable under Treat only with reverse osmosis at an annual cost in the range of $132 million. Using CmpMgmt plus treatment by coagulation a level of 10 is attainable for a cost of $9 million. For this specific level of effectiveness, then, the net savings in treatment cost from CmpMgmt is some $123 million. This amount then is an upper limit on the real costs of implementing CmpMgmt. Any implementation cost less than that represents a net cost savings.

Even though policy makers may desire to achieve a specific level of effectiveness, it is often constrained by budget limitations on what can actually be spent. It is also worthwhile, therefore, to identify the management strategy which achieves the highest level of effectiveness for a given budget. Since the cost-effectiveness curves do not incorporate strategy implementation costs, again it is necessary to examine the decision in light of allowable implementation costs. For example, if the total budget cannot exceed $15 million, then to achieve the highest level of effectiveness, CmpMgmt would be selected so long as the implementation costs were less than $4.5 million, the difference between the treatment costs for CmpMgmt ($10.5 million) and the budget constraint ($15 million) at a relative eutrophication of 10. If this criterion could not be met, then examination would proceed to other strategies.

The final question is what costs would be justified in implementing management controls for a particular pollutant source? According to the stated decision rule, this must be answered in terms of the treatment-cost savings anticipated. These cost savings, shown in Table 3, are derived from the cost-effectiveness curves of Figure 10 in the following way. The relative level of eutrophication for the lake with no management or treatment is 38. With CmpMgmt, which implements all management options, relative eutrophication is dropped 17 points to 21. At this level, the savings in treatment costs over Treat by CmpMgmt are $11.5 million. The proportion of those savings attributable to the IndDet and DomDet strategy components are assumed to be in the same proportion as the percentage of the total change in relative eutrophication that they contribute. Thus, DomDet is responsible for 12 of the total 17 points change or 70 percent. The treatment cost savings due to the DomDet strategy, then, are 70 percent of $11.5 million, or $8.1 million, and hence it pays to implement the DomDet strategy if the changeover in activities costs less than this.

Summary and Conclusions

A model incorporating phosphorus using activities, the relationship between phosphorus loading and eutrophication (Vollenweider, 1968) and the direct costs of treating phosphorus to reduce phosphorus input to surface water was developed. The model can be applied to any region or river basin using rather easily obtainable data. Using Lake Erie as an example, changes in phosphorus input as a result of various management tactics resulted in changes in relative eutrophication. Treatment costs associated with management strategies were related to effectiveness in controlling eutrophication, and the use of cost-effectiveness analysis for selecting the most comprehensive set of management factors was demonstrated. The following conclusions were generated from the study:

Table 3. Cost savings attributable to strategies for Lake Erie case.

Strategy	Relative eutrophication	Change eutrophication	Total % change for management strategy	$ savings attributable to strategy
System as is	38	--	--	
CmpMgmt	21	17	100%	$11.5 m
IndDet	35	3	18%	$ 2.0 m
DomDet	26	12	70%	$ 8.1 m
		15		
All Other		2	12%	$ 1.4 m

1. The use of activity analysis to control the effects of particular resources, in this case phosphorus, is a reasonable approach.
2. Management of a resource rather than treatment alone should be instituted when the real costs of management controls are less than the treatment costs saved or avoided.
3. In Lake Erie use of management controls can result in significant treatment cost saving and thus implementation of controls appears feasible. A combination of management controls and treatment processes can result in lowering eutrophication levels to somewhere in the mesotrophic range, at least, with respect to phosphorus loading.

References

Edmundson, W.T. 1972. Nutrients and phytoplankton in Lake Washington. *In* Nutrients and Eutrophication (G.E. Likens, Ed.). Limnology and Oceanography, Special Symposium No. 1:172-196.

Lange, W. 1971. Limiting nutrient elements in filtered Lake Erie water. Water Research 5:1031-1048.

Laurent, P. J., J. Garacher, and P. Vivier. 1970. The condition of lakes and ponds in relation to the carrying out of treatment measures. Presented at the 5th International Water Pollution Research Conference, San Francisco, California.

Lewis, R.W. 1970. Phosphorus. I No. 1970 Mineral facts and problems. U.S. Bureau of Mines. p. 1139.

Liebman, H. 1970. Biological and chemical investigations on the effect of sewage of the eutrophication of Bavarian lakes. Presented at the 5th International Water Pollution Research Conference, San Francisco, California.

Logue, P. 1958. Utilization and economics of phosphorus and its compounds. *In* Phosphorus and its Compounds (by J. Van Wazer). II:987.

Mackenthun, K.M., L.A. Leuschow, and C.D. McNabb. 1960. A study of the effects of diverting the effluent from sewage treatment upon the receiving stream. Wisconsin Academy of Sciences, Arts and Letters, Trans. 49:51-72.

Megard, R.O. 1972. Phytoplankton, photosynthesis and phosphorus in Lake Minnetonka, Minnesota. Limnology and Oceanography 17:68-87.

Oglesby, R.T. 1969. Effects of controlled nutrient dilution on the eutrophication of a lake. *In* Eutrophication: Causes, Consequences, Correctives, NAS-NRC. p. 483-493.

Porcella, D.B., A.B. Bishop, J.C. Andersen, O.W. Asplund, A.B. Crawford, W.J. Grenney, D.I. Jenkins, J.J. Jurinak, W.D. Lewis, E.J. Middlebrooks, and R.W. Walkingshaw. 1973. Comprehensive management of phosphorus water pollution. EPA-68-01-0728. Final Report to the EPA. In review. 433 p.

Porcella, D.B., P.H. McGauhey, and G.L. Dugan. 1972. Response to tertiary effluent in Indian Creek Reservoir. Journal WPCF, 44:2148-2161.

Sawyer, C.N. 1947. Fertilization of lakes by agricultural and urban drainage. Journal New England Water Works Association 61:109.

Schindler, D.W., F.A.J. Armstrong, S.K. Holmgren, and G.J. Burnskill. 1971. Eutrophication of Lake 227, experimental lakes area, Northwestern Ontario, by addition of phosphate and nitrate. Journal Fisheries Research Board, Canada 28:1763-1782.

Vollenweider, R.A. 1968. Scientific fundamentals of the eutrophication of lakes and flowing waters, with particular reference to nitrogen and phosphorus as factors in eutrophication. A Report to the Organization of Economic Cooperation and Development. Paris DAS/CSI/68. 27:1-182.

COMPONENT DESCRIPTION AND ANALYSIS OF ENVIRONMENTAL SYSTEMS: OXYGEN UTILIZATION IN AQUATIC MICROCOSMS[1]

J. Hill, IV, and D. B. Porcella[2]

Introduction

By application of increasing technological capability to the use of energy resources, man has become more important in global ecology as a force of geological magnitude than as a biological force. The negative feedback controls or homeostatic mechanisms which have evolved in ecological systems cannot cope with many of man's activities (Odum, 1971). To stem the growing environmental crisis man must become self-regulating with respect to population and industrial growth which affect the environment.

Effective self-regulation requires an operational description of the environment. The environment is a collection of biological, chemical, and physical components which, as Commoner (1971) states, "Everything is connected to everything else." A system is a set of interacting components and the environment is a system. Systems analysis, a mathematical description of system behavior, can provide an operational description of the environment which may be used to develop controls to regulate man's interaction with the environment.

The purpose of this paper is to apply the technique of component description and systems analysis to an environmental system, specifically to aquatic microcosms serving as analogues for the mud-water environment of eutrophic lakes.

Statement of the Problem

Most system analyses of large ecological or environmental systems are based upon compartment models (Walters, 1971). The principal classes of variables in these models are (1) the fluxes or rates of flow, (2) the state variables or storage levels in each compartment, (3) the inputs or forcing functions, and (4) the parameters which

multiply the state variables and inputs. There are no explicit potentials or forces which cause the flows in these models.

In an aquatic system, a potential must be measured between two points, one of which is a reference point and it is independent of the amount of material present. An example would be the chemical potential obtained by measuring dissolved oxygen in a lake and referencing to the standard state potential. Fluxes are more commonly measured and depend on the amount of material present and may be measured at a single point, for example oxygen diffusion rate from the atmosphere into water.

The potentials associated with the fluxes in models are incorporated into the parameters which modify the state variables. In the usual method of development these parameters also include many of the effects of environmental interaction with the compartment. As explained by Walters (1971), "Parameter estimates are usually found by repeatedly solving the equations, while varying the parameter estimates to obtain the best fit to the size-time data." Size-time data refer to the time series of measured outputs of the system.

These estimated parameters constitute the description of the constituents or components of the compartment. Application of these descriptive parameters to the same compartment with the same components in a different environment is not necessarily valid. For this type of extrapolation to be valid, the effects of potentials and environmental interactions must be extracted from the descriptive parameters. This is the role and special advantage of component analysis.

Explicit inclusion of potentials in models of environmental systems will allow a more realistic and useful representation of the system by (1) differentiating between potential energy storage, kinetic energy storage, and dissipation within a compartment, (2) explicitly representing the environmental resistance to flow between compartments, and (3) permitting application of the descriptive parameters for a compartment to a model of a different environment. The probability of finding two compartments (e.g., the first trophic levels in a food web) in different environmental systems with the same com-

[1] Work described herein was supported by the Office of Research and Monitoring, U.S. Environmental Protection Agency, Washington, D.C. 20460.

[2] J. Hill, IV, is with the University of Georgia, and D.B. Porcella is with the Utah Water Research Laboratory, Utah State University, Logan, Utah.

ponents (e.g., same species composition) is low. However, when potentials are included explicitly in the modeling process, the individual species may be modeled as components and then the resulting descriptive parameters may be used in any environment.

Component description of biological entities when combined with environmental potentials, fluxes, and components, provides an operational description of environmental systems which may be used to evaluate the effect of man's interaction with the environment. (For a more complete description see Hill, 1973.) This evaluation in turn may be used for guiding the self-regulation of man's activities.

In approaching any problem, an investigator must first form a mental image or conceptual model of the problem. This conceptual model is not well defined and probably varies considerably from one investigator to another. With the complex problems associated with environmental systems, solving and/or communicating the conceptual model requires translating it into the non-empirical language of mathematics or symbolic logic. However, the substance of the non-empirical representation is dependent upon the original conceptual model. In order to apply component description and analysis to environmental systems the conceptual model must include complementary variables, potentials, and fluxes.

Complementary variables, a potential, and a related flux must be included in the conceptual model of the system in order to formulate a component description of the system in the non-empirical (symbolic or mathematical) domain. An acceptable set of complementary variables is available from the fundamental equation of classical thermodynamics

$$dU = TdS - PdV + edq + Fdx + A_i df_i + \ldots \quad . \quad (1)$$

in which U is internal energy, T is temperature, s is entropy, P is pressure, V is volume, e is electrical potential, q is electrical charge, F is force, x is distance, A is chemical potential, and f is rate of change of the extent of reaction or chemical flux. However, classical application of these complementary variables to some of the processes in environmental systems is inadequate due to the non-equilibrium state of these systems. Irreversible thermodynamics provides a powerful tool in the form of cross coupling of potentials and fluxes. This allows extension of the use of these complementary variables to the description of environmental systems (Morowitz, 1968; Katchalsky and Curran, 1965).

A set of three basic components is defined for each energy form in the system. These are a resistance which represents an energy dissipation, a capacitance which represents a storage of potential energy, and an inertance which represents a storage of kinetic energy.

It is not necessary to assume linear components in the description. In fact, environmental systems exhibit inherently non-linear behavior (Fox, 1971). Non-linear components complicate the mathematical manipulation and as stated by Fox (1971), "Currently, a non-linear theory of non-equilibrium thermodynamic processes does not exist." Consequently, as a first approximation, linear components are assumed. The resulting formulation is valid if the system is not too far removed from equilibrium (Boudart, 1968; DeGroot, 1963; Callen, 1960; Prigogine, 1961). Essential non-linearities can be introduced after the linear description has been satisfactorily completed.

In addition to the three basic components which are characteristic of each energy form, there are two connective components and two energy transfer components which are universal with respect to energy. The connective components, the potential junction and the flux junction, describe the energy pathways for the interaction of the components within an energy form. The energy transfer components, the transformer, and the gyrator, describe the energy pathways for interaction between components in different energy forms.

When information is transferred with negligible flow of energy (e.g., a virus controlling a cell or a signal controlling a fluid valve), it is necessary to introduce a modulated transfer component or a controlled source, depending respectively upon whether the controlled energy is supplied by the system or the environment.

The initial component description of the system is best formulated in a graphic symbolism. The bond graph representation of Karnopp and Rosenberg (1968) is a good choice for systems in which energy flow and transformation are of primary concern. The nature of the graphic description is dependent upon the investigator's conceptual model of the processes occurring in the system and the degree of resolution desired from the resulting mathematical model. This is also related to the data which are available or which can be measured from experiments with the system. The graphical representation is the heart of the component description of the system because the applicability of ensuing analysis is limited by the ability of the investigator to represent his conceptual model of the processes in graphic form.

It should be noted here that the graphical representation can be improved by iteration. After application of the analytical techniques, any unusual or unexpected component values may be interpreted in terms of modification of the components or connectivity of the original graphical representation.

After the graphical representation is complete the system equations may be reduced to a standard form by application of state space formulation techniques. The state space formulation has the advantages of a large body of descriptive literature (e.g., Desoer, 1970; Martens and Allen, 1969; DeRusso, Roy, and Close, 1965), wide application, and extension to non-linear components. The system equations may be reduced to a state space form by algebraic or matrix manipulation (Karnopp and Rosenberg, 1968).

The system transfer function can be derived by algebraic or matrix manipulation from the state space formulation of the system equations. This transfer function, which is defined as the ratio of the Laplace transform of the output variable to that of the input variable, is a function of the component values.

The system transfer function can also be obtained from the experimental measurements of the inputs and outputs. This transfer function is expressed as an infinite series function of the complex or Laplace variable, s, (Ba Hli, 1971). By equating coefficients of equal powers of s in the two expressions for the transfer function, the component values can be determined (whether they exist as discrete physical entities or not).

The component values and state space formulation constitute a complete component description of the system in the non-empirical domain. The analysis of response, stability, and sensitivity of the system model can be pursued using differential calculus and/or computer simulation techniques which are applicable to systems of first order, linear, differential, equations with constant coefficients.

Description of Components

A component is a mathematical model of a physical process involving energy flow or transformation. The mathematical model relating the potential and flux variables for a component may be either linear or non-linear in form. Linear components are used as approximations of non-linear processes because of the greatly simplified nature of the mathematics and stability of computer simulations of the system behavior. The error due to the linear approximation is not too large if the system state does not vary too far from the state for which the approximation is made and if the state for which the approximation is made is not too far removed from equilibrium (Boudart, 1968; DeGroot, 1963; Callen, 1960; Prigogine, 1961).

The resistance, capacitance, and inertance components

The resistance is a component which is used to represent the ratio of potential difference to flow rate in a given medium. The parameter used to describe the property of the resistance component is also called the resistance. This parameter is a measure of the potential difference necessary to move a unit of flow through the component in a unit time. The product of the potential

difference and the flux associated with the resistance component is a measure of the power dissipated by the component. The mathematical representation of the linear resistance component is

$$E = RI \quad . . . . (2)$$

in which E is a potential, I is a flux and R is the resistance.

The capacitance is a component which is used to represent the storage of potential energy. The parameter used to describe the property of the capacitance component is also called the capacitance. A water tank which stores hydraulic pressure can be modeled as a fluid capacitance. The mathematical representation of the linear capacitance component is

$$I = C \left(\frac{dE}{dt} \right) . . . . (3)$$

in which C is the capacitance value and t is time.

The inertance is a component which is used to represent the storage of kinetic energy. The parameter used to describe the property of the inertance component is also called the inertance. The energy of flow which is stored in the inertia of the mass of fluid moving through a pipe can be modeled as a fluid inertance. It is this stored energy which gives rise to the water hammer effect when a valve is closed quickly. The mathematical representation of the linear inertance component is

$$E = L \left(\frac{dI}{dt} \right) . . . . (4)$$

in which L is the inertance.

Sources

Sources are components which are used to represent potentials and fluxes which originate outside the defined system boundaries. The potential source provides a potential as defined by a specified function of time. The flux source provides a flux as defined by a specified function of time. Agricultural runoff can be modeled as a chemical potential source in describing an aquatic environment.

Controlled sources are components which are used to represent coupled potentials and fluxes or information transfer. The potential or flux of a controlled source depends upon the value of a specified variable some place in the system. For example, when the flux of a chemical compound depends upon the flux of the fluid in which it is suspended or dissolved, then a controlled chemical flux source, which is dependent upon the flux value in the hydraulic sybsystem, can be included in the chemical subsystem. When a signal controls an energy process with negligible power transfer, then a form of information transfer occurs which can be modeled with a controlled

189

source. The effect of a catalyst on a chemical reaction is an example of this type of control.

Transfer components

The transformer and gyrator are transfer components which are used to represent an exchange between kinetic and potential energy within or between energy forms. The power into a transfer component equals the power out but the ratio of potential to flux for the power input differs from the ratio for the power output. For the transformer (TF) the potential in is related to the potential out and the flux in is related to the flux out. The mathematical representation of the ideal, linear transformer component is

$$E_{in} = n\ E_{out} \quad \cdots \cdots \cdots \cdots (5)$$

$$I_{in} = \frac{1}{n}\ I_{out} \quad \cdots \cdots \cdots \cdots (6)$$

in which n is the transformer ratio.

For the gyrator (GY) the potential in is related to the flux out and the flux in is related to the potential out. The mathematical representation of the ideal linear gyrator component is

$$E_{in} = m\ I_{out} \quad \cdots \cdots \cdots \cdots (7)$$

$$I_{in} = \frac{1}{m}\ E_{out} \cdots \cdots \cdots \cdots (8)$$

in which m is the gyrator ratio. A mechanical lever is an example of a transformer component. A chloroplast is an example of a gyrator component.

The modulated transformer and modulated gyrator are transfer components in which the transfer ratios (n and m) are functions of specified variables within the system. These components can be used to represent control of energy processes by information transfer.

Connective components

The potential junction and flux junction are connective components which are used to represent the pathways for interaction of components. The mathematical representation of the potential junction is

$$E_1 = E_2 = E_3 \quad \cdots \cdots \cdots \cdots (9)$$

$$I_1 + I_2 + I_3 = 0 \quad \cdots \cdots \cdots (10)$$

The mathematical representation of the flux junction is

$$E_1 + E_2 + E_3 = 0 \quad \cdots \cdots \cdots (11)$$

$$I_1 = I_2 = I_3 \quad \cdots \cdots \cdots \cdots (12)$$

The potential junction and flux junction represent parallel and series pathways, respectively.

The bond graphs of Karnopp and Rosenberg (1968) are excellent symbolic representations for component description of environmental systems where energy flow is an important process. In the bond graph representation of a system, a line segment is called a bond and represents a pathway for energy interaction of components. A dotted line segment, called an active bond, represents a pathway for information transfer.

The components are represented by circles with a mnemonic identification inside. The potential associated with a particular pathway (bond) is indicated symbolically above or on the left of the line segment. The associated flux is indicated symbolically below or to the right. An arrow on the bond indicates the direction of positive energy flow (or control for the active bond).

The causal stroke, a bar on the end of a line segment, indicates that the flux on that bond is defined by the component nearest the causal stroke and the potential by the component farthest from the causal stroke. The causal stroke defines the independent and dependent variables for each component.

These basic concepts of bond graph symbolism are summarized in Figure 1. Bond graph representation of the components is summarized in Table 1.

Graphical representation of the conceptual system model is the most important single step in the analysis of system behavior. The system equations which will be solved to determine system behavior, stability, and sensitivity are derived explicitly from the graphical representation. Decomposition of the real world into a set of interconnected components is a trial and error process based upon the investigator's conceptual model of the physical reality and the intended use of the resulting system representation. There is no best or correct graphical representation of a system, only degrees of realism.

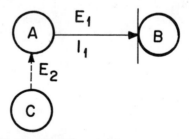

Energy flows from component A to B
E_1 is determined by component A.
I_1 is determined by component B.
I_2 is component C controls component A's function.

Figure 1. Bond graph symbolism.

"The state of a system is (defined by) the set of variables, the state variables, which contain sufficient information about the present condition of the system to permit the determination of all future time history of the system, provided that all future inputs are known." (Martens and Allen, 1969, p. 71). The potential variables on the potential energy storage components and the flux variables on the kinetic energy flux storage components usually constitute a sufficient set of state variables (Martens and Allen, 1969). The state variables define the system's "memory." In other words, the state variables are sufficient to describe the energy available in the system which may affect system behavior.

After the system has been represented in bond graph form, a systematic technique may be used to derive the system equations in a standard form. The standard form for state space equations is

$$\frac{d}{dt} [X] = [A][X] + [B][U]$$

$$[Y] = [C][X] + [D][U] \qquad \cdots \cdots \cdots (13)$$

in which the brackets indicate matrices, X is a state variable, U is an input, Y is an output, and A, B, C, and D are coefficients.

Table 1. Component bond graphs and mathematical representations.

Component	Bond Graph	Mathematical Representation
Resistance		$E = RI$
Capacitance		$I = C\, dE/dt$
Inertance		$E = L\, dI/dt$
Potential Source		$E_1 = E(t)$
Flux Source		$I_1 = I(t)$
Transformer		$E_1 = nE_2$ $I_1 = (1/n)I_2$
Gyrator		$E_1 = mI_2$ $I_1 = (1/m)E_2$
Potential Junction		$E_1 = E_2 = E_3$ $I_1 + I_2 + I_3 = 0$
Flux Junction		$E_1 + E_2 + E_3 = 0$ $I_1 = I_2 = I_3$

The first step in systematic reduction of the bond graph to the state space equations is to name the bonds. A numerical subscript on the pair of variables on each bond serves to identify each component and variable.

The next step is to assign causality to each bond in the graph. This defines the independent variable in each of the component equations. The following steps are used to assign causality to the bonds in the graph.

1. The potential on the bond of a potential source is an independent variable. The potential on this bond is defined or "caused" by the potential source.
2. The flux on the bond of a flux source is an independent variable.
3. The potential on the bond of a capacitance is an independent variable.
4. The flux on the bond of an inertance is an independent variable.
5. The independent variables determined by steps 1 through 4 are used in the equations for the connective components (Equations 9 through 12) to define the independent variables on the connective component bonds.
6. The independent variables on the bonds of the resistances are chosen so that all bonds have a causal stroke which defines the independent variables.

Conflicts in causal assignment may indicate an unrealistic graphical representation of the physical situation.

At this stage the component equations may be transcribed and algebraically manipulated to the standard state space form (Equations 13). However, for large systems (greater than three storage components) the algebra becomes awkward and the matrix reduction which follows is recommended.

To continue the systematic reduction, the bond graph variables are divided into five classes. The state variables (X) are the potentials on capacitances and the fluxes on inertances. The input variables (U) are the potentials on potential sources and the fluxes on flux sources. The temporary variables (T) are the independent variables on the resistances. The auxiliary variables (H) are the independent variables on the transfer components (TF and GY) and the connective components (JP and JF). The output variables (Y) are chosen by the investigator for the descriptive purposes.

The relations for a system of linear components may be written in matrix form as

$$d/dt[X] = [C_{11}][X] + [C_{12}][T] + [C_{13}][H] + [C_{14}][U]$$

$$[T] = [C_{21}][X] + [C_{22}][T] + [C_{23}][H] + [C_{24}][U]$$

$$[H] = [C_{31}][X] + [C_{32}][T] + [C_{33}][H] + [C_{34}][U]$$

$$\dots\dots\dots\dots (14)$$

Equations 14 may be reduced to standard state space form in the following manner:

$$[H] = ([I_d] - [C_{33}])^{-1} ([C_{31}][X] + [C_{32}][T] + [C_{34}][U])$$

$$[H] = [C'_{31}][X] + [C'_{32}][T] + [C'_{34}][U]$$

$$[T] = \{[I_d] - ([C_{22}] + [C_{23}][C'_{32}])\}^{-1} \{([C_{21}]$$
$$+ [C_{23}][C'_{31}])[X] + ([C_{24}] + [C_{23}][C'_{34}])[U]\}$$

$$[T] = [C'_{21}][X] + [C'_{24}][U]$$

$$d/dt[X] = ([C_{11}] + [C_{12}][C'_{21}] + [C_{13}][C'_{31}]$$
$$+ [C_{13}][C'_{32}][C'_{21}])[X] + ([C_{14}] + [C_{12}][C'_{24}]$$
$$+ [C_{13}][C'_{34}] + [C_{13}][C'_{32}][C'_{24}])[U]$$

$$d/dt[X] = [C''_{11}][X] + [C''_{14}][U] \qquad \dots\dots (15)$$

Equations 15 are of the same form as the standard state space Equations 13 in which $[A] = [C''_{11}]$ and $[B] = [C''_{14}]$.

System Transfer Function

The system function or transfer function (H(s)) is defined for a system without time varying coefficients as the ratio of the Laplace transform of the output to the Laplace transform of the input when there is no initial stored energy (DeRusso et al., 1965).

$$H(s) = Y(s)/U(s) \qquad \dots\dots\dots\dots (16)$$

A time varying transfer function (H(t,s)) may be found for systems with time varying coefficients (DeRusso et al., 1965). A matrix transfer function ([H(s)]) may be determined for systems with more than one input and/or output (Ogata, 1967).

It is useful in the analysis of environmental systems to amend the definition of the transfer function to allow non-zero values of initial stored energy since some environmental systems cease to function when internal energy equals zero. In this case the observed transfer function ($H^*_{ab}(s)$) between output a and input b becomes

$$H^*_{ab}(s) = [C_{aj}](s[Id] - [A])^{-1}[B_{jb}] + [D_{ab}]$$
$$+ \{[C_{aj}](s[Id] - [A])^{-1}[X(0)]\}/[U_b(s)] . (17)$$

in which [A], [B], [C], and [D] are the coefficient matrices of the state equations, Id is the identity matrix, j is the number of state variables, [X(0)] is the initial value vector, and s is the complex frequency or Laplace variable. The observable transfer function of the system for a given input may be obtained from the bond graph through the system equations by application of Equation 17.

The observable transfer function may also be determined from the input and output data which are obtained from system measurements (Ba Hli, 1971). If the input is expressed as a time series ($\{U_1, U_2, U_3, ...\} = \{U\}$) and the output is expressed as a time series ($\{Y_1, Y_2, Y_3, ...\} = \{Y\}$), then the time series of areas under the impulse response curve $\{h_a\}$ can be found by synthetic division of the output time series by the input time series as shown in Figure 2. The observable transfer function $H^*(s)$ may be determined from $\{h_a\}$ by

$$H^*(s) = \sum_1^n h_{an} - s\Sigma h_{an} t_n + \frac{s^2}{2!} \Sigma h_{an} t_n^2 - \quad . . (18)$$

in which s is the complex frequency or Laplace variable, h_{an} is the value of h_a in the nth time interval, and t_n is the time value at the middle of the nth time interval.

$$U_1, U_2, U_3, \cdots \frac{h_{a1}, h_{a2}, h_{a3}, \cdots h_{an}}{Y_1, Y_2, Y_3, \cdots\cdots Y_n}$$
$$\frac{a_1, a_2, a_3, \cdots\cdots a_n}{b_2, b_3, \cdots\cdots b_n}$$
$$\frac{c_2, c_3, \cdots\cdots c_n}{}$$

Figure 2. Synthetic division.

Component Analysis

Some of the energy processes in environmental systems which may be modeled as components are not physically discrete entities. The chemical resistance component in a representation of the processing of oxygen by a species of algae cannot be placed on a table and subjected to varying oxygen affinities to determine the value of its resistance parameter. However, values of some of the components may be found by using the observable transfer function ($H^*(s)$).

The observable transfer function is expressed as a function of component values by Equation 17. It is also expressed as a function of the input-output data by Equation 18. By equating the coefficients of like powers of s in these two expressions, j + 1 independent equations may be generated in which j is the number of state variables. These equations may generally be solved for j + 1 unknown component values.

The component values other than the j + 1 accounted for above must be found by other means. Some component values may be easily measured in the physical system and others by specially designed experiments. If the system exhibits a steady state, it is possible in some

cases to determine additional component values by setting the derivatives in the state equations to zero.

System Analysis

The state space equations and component values constitute an operational description of the environmental system within the limits imposed by the assumptions and approximations which are made. This mathematical description or model may be used with analytical or simulation techniques to approximate the response of the physical system to modified inputs, damaged components, and other phenomena. It is certainly more reasonable to use a mathematical model to investigate approximate responses to environmental changes than to wait and see what happens to the environment when it is perturbed, e.g. by pollution.

In addition to mimicing the physical system behavior, the state equation model may be used to evaluate properties of the system as a whole (emergent properties). These mathematically derived properties are related to the energy processing and control structure of the system. They would be difficult, if not impossible, to determine by experiment on environmental systems, which cannot generally be stopped and restarted in a specified initial state. These total system properties are considered in the context of general system theory (Bertalanffy, 1968).

System Properties

The property of stability is a measure of the "boundedness" of the system response as time approaches infinity (DeRusso et al., 1965). In other words, does the system approach a steady state? The property of controllability is a measure of the degree to which the state or output of the system may be modified by inputs to produce a specified state or output within a finite time (Ogata, 1967). The property of observability is a measure of the degree to which the state of the system may be determined from a knowledge of the output over a finite time interval (Ogata, 1967). The property of optimality is a measure of how well a system meets a defined performance index (DeRusso et al., 1965). The stability, controllability, observability, and optimality of linear, time varying, and non-linear systems is treated extensively in the system analysis literature.

The property of sensitivity is a measure of the state or output change resulting from a parameter perturbation. Sensitivity to a state perturbation from outside the system may also be postulated. Sensitivity measures may also be used to evaluate system response to inherent errors in parameter or initial state determination (Astor et al., 1972). There is a possibility that in highly connected environmental systems the sensitivity along energy processing pathways may be reciprocally related to the sensitivity along information transferring pathways (Patten, 1972). Sensitivity analysis of environmental systems is relatively new and may provide some valuable general statements concerning information and energy in environmental control processes.

The property of independence or summativity is a measure of the degree to which total system response is independent of component interaction (Bertalanffy, 1968). A related property, centralization, or individualization, is a measure of the degree to which total system response is dependent upon a single component or group of components (Bertalanffy, 1968). These two total system properties may become important in describing the evolution of environmental systems; however, their current usage appears to be more philosophical than mathematical.

Component Description and Analysis of Oxygen in an Aquatic Microcosm

System description

The aquatic microcosms shown in Figure 3 (adapted from Porcella et al., 1970) are part of a sediment-water nutrient exchange experiment being conducted at the Utah Water Research Laboratory, Logan, Utah. The sediment samples were taken from Hyrum Reservoir, Utah (see Porcella et al., 1972).

The microcosms consisted of approximately 75 cm high lucite cylinders which had an inside diameter of 14 cm. They were isolated from the atmosphere by a gas-trap which had a provision for removal of gas samples. The microcosms were filled to a depth of approximately 15 cm (2.3 liters) with sediments and then with water to a level within 3 cm of the top seal.

There were 16 microcosms in the experiment arranged in a 4 x 2 x 2 factorial experimental design. The variations of treatments were 1) light (dark, vertical continuous light, horizontal diurnal light of 16 hours, and horizontal light with variable intensity over a diurnal cycle); 2) nitrogen (high and low levels); and 3) mercury (high and low levels). The particular microcosm selected for study in this part is number nine which has vertical continuous light, low nitrogen, and low mercury treatments. Only one microcosm was selected to demonstrate this technique; this was microcosm number nine and it was selected because it has demonstrated considerable suspended growth of both algae and bacteria and gas production (mostly O_2) compared to the other microcosms.

Each day, 10 percent (about .9 liter) of the volume of water is removed and replaced with fresh nutrient media. Thus the water in the microcosms has a mean residence time of about 10 days.

The water in the microcosms is completely mixed with a water driven magnetic stirrer, and maintained at a temperature of approximately 25°C.

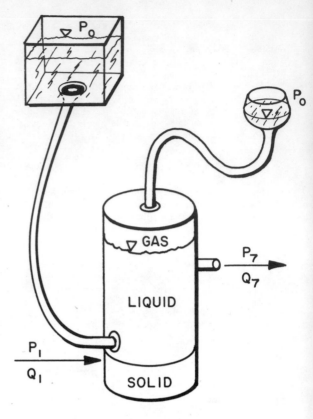

Figure 3. Pictorial diagram of the aquatic microcosm hydraulic subsystem.

A complete description of the microcosms, experimental techniques, data, and results can be found in the Utah Water Research Laboratory project report on OWRR Project No. B-081 (Porcella et al., 1973).

The bond graph

The energy processes associated with oxygen in microcosm number nine (henceforth called the microcosm) are to be studied using a component description and analysis of the oxygen subsystem. The possible component representations of the oxygen subsystem in the microcosm are innumerable. The representation which follows (Figure 4) is a compromise between simplicity and realism.

In the bond graph of Figure 4, f is a chemical flux (mg/day), A is a chemical potential (Kcal/mg), C is a capacitance for the storage of chemical potential energy, R is a chemical resistance, and TF is a transformer. The component indicated by an arrow is a unilateral flux component. It indicates that flow can only occur in the direction indicated.

The causality and resulting independent variables are indicated on the bonds of the bond graph in Figure 4. The other variables on the bonds which result from the use of the independent variables in the junction component equations appear in Figure 5.

f_1 is the input of dissolved oxygen in the nutrient media. It is assumed to be a constant value of 7.56 mg/day.

C_3 represents the storage of dissolved oxygen in the water in the body of the microcosm.

R_5 represents the resistance to oxygen exchange between the gas and liquid phases.

C_6 represents the storage of gaseous oxygen in the gas-trap.

R_{10} represents the resistance to the photosynthesis driven breakdown of water to oxygen.

C_{14} represents the storage of oxygen as water.

f_{17} is a controlled flux source which represents the flux of oxygen from dissolved oxygen to the oxygen of water through respiration.

A_{20} is an hypothetical, chemical, potential, source which represents an energy available for the growth of biomass. This source includes the chemical energies of all nutrients and that which results from electromagnetic excitation of the chlorophyll molecules. If one of these energy sources were varying considerably and limiting growth, then the energy processing subsystem for that energy source would have to be coupled to the oxygen subsystem to produce a realistic model. In an ideal model A_{20} would be replaced by individual sources and energy processing components for each energy input.

R_{21} represents the growth resistance of the biomass.

C_{25} represents the storage of biomass in the microcosm.

α_{17} is the ratio of the flux, f_{17}, to the potential of the biomass (A_{25})

β_{19} is the ratio of the flux of oxygen out of the microcosm (f_{19}) to the potential of oxygen in the microcosm (A_3). f_{19} depends on the flux of water out of the microcosm (0.9 liter/day) and the potential (which is proportional to the concentration) of oxygen in the microcosm.

R_{26} represents the decay resistance of the biomass.

TF, the transformer component, represents the use of some of the energy available for photosynthesis to produce oxygen from water.

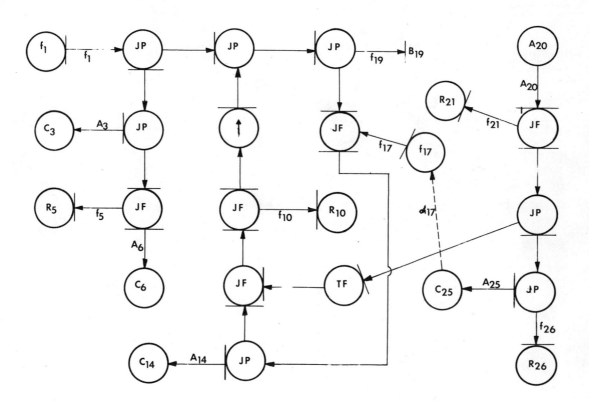

Figure 4. Bond graph of the oxygen subsystem of the microcosm with the independent variables indicated.

The component equations

The component equations may be written from the bond graph (Figure 5) by using the information in Table 1.

The state variables (X) are A_3, A_6, A_{14}, and A_{25}. The temporary variables (T) are f_5, f_{10}, f_{17}, f_{19}, f_{21}, and f_{26}. There are no auxiliary variables (H) because the potentials and fluxes on the transformer component were written in terms of state variables, temporary variables, and transformer ratio (n). The input variables (U) are f_1 and f_{20}. The output variables (Y) are the same as the state variables.

The component equations are:

$$\dot{A}_3 = C_3^{-1} (f_1 - f_5 + f_{10} - f_{17} - f_{19}) \quad \cdots \cdots (19)$$

$$\dot{A}_6 = C_6^{-1} f_5 \quad \cdots \cdots \cdots \cdots (20)$$

$$\dot{A}_{14} = C_{14}^{-1} (-f_{10} + f_{17}) \quad \cdots \cdots \cdots (21)$$

$$\dot{A}_{25} = C_{25}^{-1} (-nf_{10} + f_{21} - f_{26}) \quad \cdots \cdots (22)$$

$$f_5 = R_5^{-1} (A_3 - A_6) \quad \cdots \cdots \cdots \cdots (23)$$

$$f_{10} = R_{10}^{-1} (-A_3 + A_{14} + nA_{25}) \quad \cdots \cdots (24)$$

$$f_{17} = \alpha_{17} A_{25} \quad \cdots \cdots \cdots \cdots \cdots (25)$$

$$f_{19} = \beta_{19} A_3 \quad \cdots \cdots \cdots \cdots \cdots (26)$$

$$f_{21} = R_{21}^{-1} (A_{20} - A_{25}) \quad \cdots \cdots \cdots (27)$$

$$f_{26} = R_{26}^{-1} A_{25} \quad \cdots \cdots \cdots \cdots (28)$$

These component equations are presented in matrix form in Figure 6.

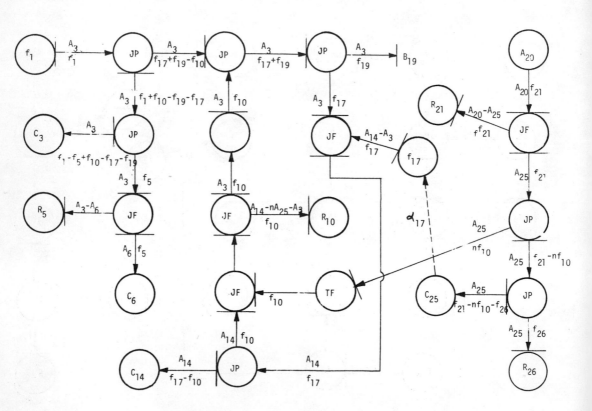

Figure 5. Completed bond graph of the oxygen subsystem.

$$\frac{d}{dt}\begin{bmatrix} A_3 \\ A_6 \\ A_{14} \\ A_{25} \end{bmatrix} = \begin{bmatrix} & & \nearrow \\ & O & \\ & & \searrow \end{bmatrix}\begin{bmatrix} A_3 \\ A_6 \\ A_{14} \\ A_{25} \end{bmatrix} + \begin{bmatrix} -C_3^{-1} & C_3^{-1} & C_3^{-1} & C_3^{-1} & 0 & 0 \\ C_6^{-1} & 0 & 0 & 0 & 0 & 0 \\ 0 & -C_{14}^{-1} & C_{14}^{-1} & 0 & 0 & 0 \\ 0 & -nC_{25}^{-1} & 0 & 0 & C_{25}^{-1} & C_{25}^{-1} \end{bmatrix}\begin{bmatrix} f_5 \\ f_{10} \\ f_{19} \\ f_{19} \\ f_{21} \\ f_{26} \end{bmatrix} + \begin{bmatrix} C_3^{-1} & 0 \\ 0 & 0 \\ 0 & 0 \\ 0 & 0 \end{bmatrix}\begin{bmatrix} f_1 \\ A_{20} \end{bmatrix}$$

$$\begin{bmatrix} f_5 \\ f_{10} \\ f_{17} \\ f_{19} \\ f_{21} \\ f_{26} \end{bmatrix} = \begin{bmatrix} R_5^{-1} & -R_5^{-1} & 0 & 0 \\ -R_{10}^{-1} & 0 & R_{10}^{-1} & nR_{10}^{-1} \\ 0 & 0 & 0 & \alpha_{17} \\ \beta_{19} & 0 & 0 & 0 \\ 0 & 0 & 0 & -R_{21}^{-1} \\ 0 & 0 & 0 & R_{26}^{-1} \end{bmatrix}\begin{bmatrix} A_3 \\ A_6 \\ A_{14} \\ A_{25} \end{bmatrix} + \begin{bmatrix} & & \nearrow \\ & O & \\ & & \searrow \end{bmatrix}\begin{bmatrix} f_5 \\ f_{10} \\ f_{17} \\ f_{19} \\ f_{21} \\ f_{26} \end{bmatrix} + \begin{bmatrix} 0 & 0 \\ 0 & 0 \\ 0 & 0 \\ 0 & 0 \\ 0 & R_{21}^{-1} \\ 0 & 0 \end{bmatrix}\begin{bmatrix} f_1 \\ A_{20} \end{bmatrix}$$

Figure 6. Matrix form of the component equations.

State Space Equations

The coefficient matrices in Figure 6 can be identified as the coefficient matrices (C_{ij}) in the general form of the matrix equations (Equations 14). The matrix equations in Figure 6 may be reduced to the following state space equations (as in Equations 13):

$$\dot{A}_3 = (-R_5^{-1}C_3^{-1} - R_{10}^{-1}C_3^{-1} - \beta_{19}C_3^{-1})A_3 + R_5^{-1}C_3^{-1}A_6$$

$$+ R_{10}^{-1}C_3^{-1}A_{14} + (nR_{10}^{-1} - \alpha_{17})C_3^{-1}A_{25} \quad \ldots (29)$$

$$\dot{A}_6 = R_5^{-1}C_6^{-1}A_3 - R_5^{-1}C_6^{-1}A_6 \quad \ldots \ldots (30)$$

$$\dot{A}_{14} = R_{10}^{-1}C_{14}^{-1}A_3 - R_{10}^{-1}C_{14}^{-1}A_{14}$$

$$- (nR_{10}^{-1} - \alpha_{17})C_{14}^{-1}A_{25} \quad \ldots \ldots (31)$$

$$\dot{A}_{25} = n^{-1}R_{10}^{-1}C_{25}^{-1} - n^{-1}R_{10}^{-1}C_{25}^{-1}A_{14}$$

$$- (n^2R_{10}^{-1}C_{25}^{-1} + R_{21}^{-1}C_{25}^{-1} - R_{26}^{-1}C_{25}^{-1})A_{25} \quad .(32)$$

These equations are a mathematical description of the energy processes in the oxygen subsystem of the microcosm as they are represented in the bond graph (Figure 5).

System Transfer Functions

The observable transfer function of the oxygen subsystem of the microcosm for each output may be obtained from the state equations of the system (Equations 29 through 32) and Equation 17. The transfer functions which result are:

$$H_3^*(s) = \frac{m}{s - a} + \frac{A_3(0)}{(s - a)U(s)} \quad \ldots \ldots (33)$$

197

$$H_6^*(s) = \frac{-bm}{(s-a)(s+b)} + \left(\frac{A_6(0)}{s+b} - \frac{bA_3(0)}{(s-a)(s+b)}\right) U(s)^{-1}$$

$$\cdots \cdots \cdots (34)$$

$$H_{14}^*(s) = \frac{p}{s+c} + \frac{mp}{(s-a)(s+c)} + \left(\frac{A_{14}(0)}{s+c}\right.$$

$$\left. + \frac{A_3(0)c}{(s-a)(s+c)}\right) U(s)^{-1} \cdots \cdots (35)$$

$$H_{25}^*(s) = \frac{g}{s-e} + \frac{pd}{(s+c)(s-e)} - \frac{md}{(s-a)(s-e)}$$

$$- \frac{mcd}{(s-a)(s+c)(s-b)} + \left(\frac{dA_{14}(0)}{(s+c)(s-e)}\right.$$

$$\left. - \frac{A_3(0)d}{(s-a)(s-e)} - \frac{A_3(0)dc}{(s-a)(s+c)(s-e)}\right) U(s)^{-1}$$

$$\cdots \cdots \cdots (36)$$

in which $A_i(0)$ is the initial value of the state variable A_i and the following substitutions of component values should be observed.

$$a = -\beta_{19} C_3^{-1} \quad \cdots \cdots \cdots \cdots (37)$$

$$b = R_5^{-1} C_6^{-1} \quad \cdots \cdots \cdots \cdots \cdots (38)$$

$$c = R_{10}^{-1} C_{14}^{-1}(-n^2 R_{10}^{-1} - R_{21}^{-1} + R_{26}^{-1}) C_{25}^{-1}$$

$$+ nR_{10}^{-1} C_{25}^{-1} C_{14}^{-1}(nR_{10}^{-1} - \alpha_{17}) \cdots \cdots (39)$$

$$d = n^{-1} R_{10}^{-1} C_{25}^{-1} \quad \cdots \cdots \cdots \cdots (40)$$

$$e = _{25}^{-1}(-n^2 R_{10}^{-1} - R_{21}^{-1} + R_{26}^{-1}) \quad \cdots \cdots (41)$$

$$g = R_{21}^{-1} C_{25}^{-1} \quad \cdots \cdots \cdots \cdots (42)$$

$$m = C_3^{-1} \quad \cdots \cdots \cdots \cdots \cdots (43)$$

$$p = R_{21}^{-1} C_{25}^{-1} C_{14}^{-1}(nR_{10}^{-1} - \alpha_{17}) \quad \cdots \cdots (44)$$

The observable transfer function for each output may also be obtained from the synthetic division of the output time series by the input time series (Figure 2) and Equation 18.

The time, mid-interval time, output, and the result of the synthetic division (h_a) for each output except water (A_{14}) which was assumed constant, are presented in Tables 2 and 3. Using the values of h_a and t from these tables in Equation 18 results in the following representations of the observable transfer functions.

$$H_3^*(s) = 2.17(10^{-5}) - 2.53(10^{-4})s \quad \cdots \cdots (45)$$

$$H_6^*(s) = 3.39(10^{-5}) - 7.95(10^{-4})s + 6.82s^2 \quad \cdot (46)$$

$$H_{14}^*(s) = 3.02(10^{-3}) - 2.11(10^{-2})s$$

$$+ 7.40(10^{-2})s^2 \quad \cdots \cdots \cdots \cdots (47)$$

$$H_{25}^*(s) = 5.32(10^{-3}) + 1.32s \quad \cdots \cdots \cdots (48)$$

Component Analysis

Some of the component values may be obtained mathematically by equating the two expressions for each transfer function (Equations 33 through 36 and 45 through 48). The other component values must be obtained either from measurement of the system or from the literature.

Equating the expression for $H_3^*(s)$ from Equation 33 with the expression for $H_3^*(s)$ from Equation 45 results in an equation with β_{19} and C_3 as the unknowns. The coefficients of like powers of s on each side of the equation must be equal so the following equations result:

$$C_3^{-1} = -2.17(10^{-5})\beta_{19} C_3^{-1} \quad \cdots \cdots (49)$$

$$\frac{(-3.02)(10^{-5})}{7.56} = (2.17)(10^{-5})$$

$$- (2.53)(10^{-4})\beta_{19} C_3^{-1} \quad \cdots \cdots (50)$$

These equations may be solved for β_{19} and C_3.

Performing the same operations with the expressions for $H_6^*(s)$ gives:

$$\beta_{19} = -4.94(10^4) \quad \cdots \cdots \cdots \cdots (51)$$

$$R_5^{-1} C_6^{-1} = 7.46 \, (10^{-2}) \quad \ldots \ldots \ldots (52)$$

Equation 52 can be used to determine either R_5 or C_6 once the other is known.

Equating the expressions for $H_{14}^*(s)$ and separating the equations for like coefficients results in two equations with n, R_{10}, C_{14}, α_{17}, R_{21}, C_{25}, and R_{26} as unknowns. Any two of the unknowns may be determined when the rest are known.

Table 2. Output data for dissolved oxygen (A_3) for the oxygen subsystem of the microcosm.

Time (Days) =	Mid-interval Time (Days) =	Dissolved Oxygen Output (A_{32} mg/l) =	Synthetic Division Quotient h_a (mg/l)
28	14.0	12.8	12.8
36	32.0	14.7	1.9
43	39.5	15.0	0.3
50	46.5	14.6	-0.4
57	53.5	12.4	-2.2
64	60.5	16.2	3.7
71	67.5	13.6	-2.5
78	74.5	14.1	0.5
85	81.5	13.3	-0.8
92	88.5	14.5	1.2
99	95.5	13.5	-1.0
107	103.0	12.7	-0.8
113	109.0	13.9	1.2
120	116.5	11.9	-2.0
127	123.5	12.0	0.1
134	130.5	12.9	0.9
141	137.5	14.0	1.1
148	144.5	15.3	1.3
155	151.5	13.5	-1.8
162	158.5	12.8	-0.7

Table 3. Output data for oxygen gas (A_6) for the oxygen subsystem of the microcosm.

Time (Days) =	Mid-interval Time (Days) =	Oxygen Gas Output (% O_2) =	Synthetic Division Quotient h_a (% O_2) =
7	3.5	19.8	19.8
14	10.5	25.5	5.7
28	21.0	45.1	19.6
43	35.5	39.2	-5.9
56	49.0	41.0	1.8
70	63.0	36.4	-4.6
84	77.0	36.4	0.0
98	91.0	42.2	5.8
111	104.5	41.4	-0.8
126	119.5	31.7	-9.7
140	133.0	40.3	8.6
154	147.0	43.0	2.7
168	161.0	43.0	0.0
182	175.0	43.0	0.0

Finally, performing the same operations for $H_{25}^*(s)$ results with two equations with R_5, C_6, n, R_{10}, C_{14}, α_{17}, R_{21}, C_{25}, and R_{26} as unknowns. These two equations may be solved for two unknowns if the rest have been evaluated by other means.

The capacitance values (C_3, C_6, C_{14}, and C_{25}) may be determined by analysis of the physical and chemical aspects of the processes of interest in the microcosms. The capacitance is related to the potential and the flux by the basic linear component equation,

$$I = C \frac{dE}{dt} \quad \text{or} \quad f = C \frac{dA}{dt} \quad \ldots \ldots \ldots (53)$$

Equation 53 indicates that the capacitance is equal to the reciprocal of the rate of change of potential per unit flux. The chemical potential (A) is related to the concentration of the chemical species ([Z]) by

$$A = A_0 + RT \ln [Z] \quad \ldots \ldots \ldots \ldots (54)$$

in which A_0 is the potential in a defined reference state, R is the universal gas constant, and T is the absolute temperature. The change in concentration is related to the flux and the volume over which the flux is distributed. Thus the rate of change of chemical potential is a logrithmic function of the flux and C must be based upon a linear approximation. If the rate of change of the potential is evaluated using Equation 54 for a flux of 1 mg/day with an initial concentration midway between the extremes measured on the system, then an approximate linear capacitance value may be determined. The approximate values are presented in Table 4.

One more component value is necessary in order to solve the equations resulting from $H_{14}^*(s)$ and $H_{25}^*(s)$ for the rest of the component values.

Table 4. Reference potentials and capacitance values for the oxygen subsystem of the microcosm.

Storage of	Component	$A_0(\frac{Kcal}{mole})$	Capacitance $(\frac{mg^2}{Kcal})$
Dissolved O_2	C_3	3.9[a]	5.60×10^6
Gaseous O_2	C_6	0.0[b]	2.23×10^6
Water	C_{14}	-56.7[b]	2.98×10^9
Biomass	C_{25}	599.4[c]	5.95×10^3

[a]Wagman et al., 1965.
[b]Weast, 1970.
[c]Morowitz, 1968.

α_{17} is the ration of the flux of oxygen from dissolved oxygen into oxygen of water to the biomass potential. It is then the oxygen respiration rate. An average value of 1.13 (10^4) mg^2/Kcal-day at $25°C$ is assumed (DiToro et al., 1971).

The measured and derived component values are presented in Table 5.

System Analysis

The set of state space system equations (Equations 29 through 32) along with the component values constitutes a complete dynamic description of the system. These equations may be solved for the outputs as a function of time by analytical, analog computer, or digital computer techniques. The equations were solved with a digital computer program using the Mimic Simulation language (see Stephenson, 1971).

The measured outputs and the result of the computer solution of the system equations are presented in Figures 7 and 8.

Improvement of the oxygen model

The application of component analysis to the oxygen subsystem gives an indication of the difficulties encountered when many energy processes are involved in the physical process. The accuracy of the model of the oxygen subsystem may be improved in many ways. One way would be to include some of the observed behavior of the system as inputs to the model. For example, the organic carbon concentration might be used to generate values for a time varying resistance to control the flux of oxygen caused by photosynthesis and respiration.

A further extension of the preceding improvement might be to use the observed values of organic carbon as a time varying coefficient of a component to control oxygen flux due to respiration. In the same model, the observed values of relative fluorescence (or some other estimate of algal biomass) may be used as an input to controlled oxygen flux source representing oxygen flux due to photosynthesis.

Each of the above improvements would probably increase the value of the model as a tool for prediction of system behavior. However, neither of the proposed improvements would increase the accuracy of the component values or the level of physical reality in the model. The measured data which are inputs to the model may improve simulation but they add nothing to understanding the processes which are modeled as components.

Enlarging component models

Another method of improving the model of the oxygen subsystem is to include more of the energy

Table 5. Measured and derived component and parameter values for the oxygen subsystem of the microcosm.

Component or Parameter	Measured Value	Derived Value
C_3	$5.60 \, (10^6) \, mg^2/Kcal$	$4.52 \, (10^5) \, mg \, /Kcal$
C_6	$3.23 \, (10^6) \, mg^2/Kcal$	-
C_{14}	$2.98 \, (10^9) \, mg^2/Kcal$	-
C_{25}	$5.95 \, (10^3) \, mg^2/Kcal$	-
R_5	-	$4.15 \, (10^{-6}) \, Kcal\text{-}day/mg^2$
R_{10}	-	$2.35 \, (10^{-6}) \, Kcal\text{-}day/mg^2$
R_{21}	-	$2.37 \, (10^{-1}) \, Kcal\text{-}day/mg^2$
R_{26}	-	$2.41 \, (10^{-3}) \, Kcal\text{-}day/mg^2$
α_{17}	$1.13 \, (10^4) \, mg^2/Kcal\text{-}day$	-
β_{19}	-	$4.60 \, (10^4) \, mg^2/Kcal\text{-}day$
$A_3(0)$	$-3.02 \, (10^{-5}) \, Kcal/mg$	-
$A_6(0)$	$-2.98 \, (10^{-5}) \, Kcal/mg$	-
$A_{14}(0)$	$-3.02 \, (10^{-3}) \, Kcal/mg$	-
$A_{25}(0)$	$0.0 \, Kcal/mg$	-
f_1	$7.56 \, mg/day$	-
n	$5.80 \, (10^{-2})$	-

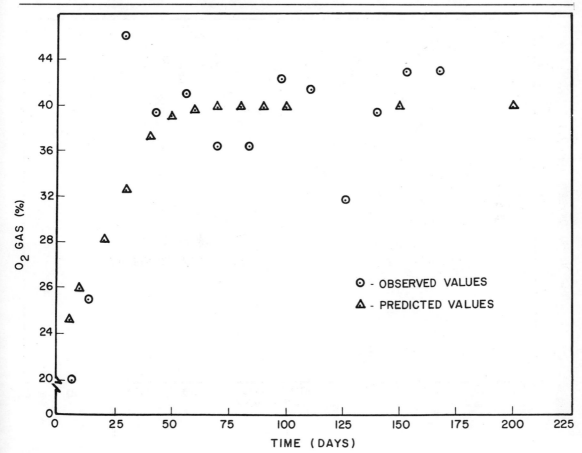

Figure 7. Oxygen gas (%) versus time (days).

processes which occur in the physical system in the model. This involves redrawing the system bond graph and starting again from the component equations. The first model is an approximation and shows that the technique is feasible. Also several of the component values will not change in a more complete representation of the oxygen subsystem. To improve the model further, light input and a separation of autotrophs and heterotrophs could be applied. The decay rate, growth rate, respiration rate and photosynthetic rates of autotrophic and heterotrophic population could be isolated and represented by separate components.

Continuation of these improvements to include components which represent individual species and coupling of the oxygen subsystem to the carbon dioxide or nitrogen subsystems is limited only by the time and effort necessary to complete the analysis and make the necessary measurements.

Advantages and disadvantages of component description and analysis

A major advantage of model improvement by addition of components or decomposition of previous components is that when a component value is determined for an isolated process then that component value is applicable to the same component in another model. Conversely, a major disadvantage of enlarged component models is the complexity of the mathematics involved in the component analysis.

An important advantage of component modeling and analysis occurs when performed before the final design of an experiment. Prior to beginning a series of measurements and experiments on a biological or environmental system an investigator can determine which measurements are important or necessary to complete the objectives of the study.

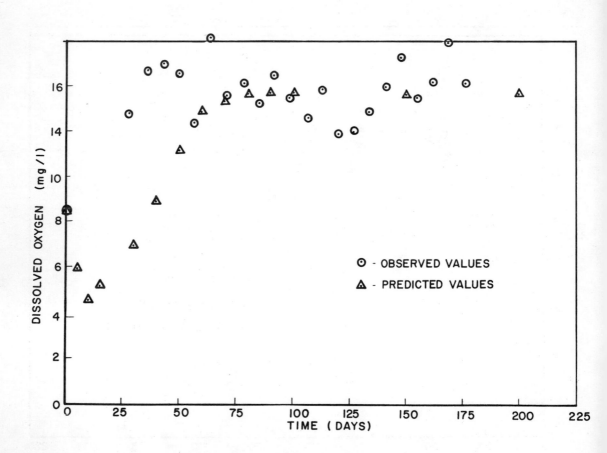

Figure 8. Dissolved oxygen (mg/l) versus time (days).

A cursory component analysis with a simple hypothesized graph structure can help to determine the minimum data necessary to characterize the system. In order of magnitude estimates are made for component values, then a simple sensitivity analysis will indicate which measurements are most critical in characterizing the system. Finally, time constants may be approximated from the estimated component values and structure and the frequency of measurements may be estimated.

General Conclusions

Component description of environmental systems is a detailed method of arriving at a system model which includes a great deal of information concerning processes of energy transfer and transformation. If component values are not available or accessible for measurement, the component analysis is tedious and expensive. However, the component values resulting from a component analysis are applicable to the same component in a similar system; therefore, some component analyses need only be done once.

Modeling of environmental systems for impact studies, development planning, or similar purposes with specific applications to a specific area is probably best done using compartmental techniques. Modeling of environmental systems for understanding of processes, general application of controls, and man-made perturbations is probably best done using component techniques. The analysis demonstrated in this paper illustrates the analytical process and attempts to point out some of the pitfalls as well as advantages in the application of the technique.

References

Astor, P.H., B.C. Patten, and J. Estberg. 1972. On sensitivity of ecosystem models. In Project Reports: 1972 Summer Institute on the Systems Approach to the Life Sciences. Utah State University, Logan, Utah.

Ba Illi, F. 1971. A time domain approach. In Aspects of Network and System Theory. Kalman, R.E., and N. DeClaris, ed. Holt, Rinehart, and Winston, New York, p. 313-325.

Bertalanffy, L. von. 1968. General systems theory. George Braziller, Inc., New York. 289 p.

Boudart, M. 1968. Kinetics of chemical processes. Prentice-Hall, Inc., Englewood Cliffs. 246 p.

Callen, H. B. 1960. Thermodynamics. John Wiley & Sons, New York.

Commoner, B. 1971. The closing circle. Alfred A. Knopf, Inc., New York. 326 p.

DeGroot, S.R. 1963. Thermodynamics of irreversible processes. North-Holland Publishing Co., Amsterdam. 242 p.

DeRusso, R.M., R. Roy, and C. Close. 1965. State variables for engineers. John Wiley & Sons, New York.

Desoer, C.A. 1970. Notes for a second course on linear systems. Van Nostrand, Reinhold, New York. 199 p.

DiToro, D.M., D.J. O'Connor, and R.V. Thomann. 1971. A dynamic model of the phytoplankton population in the Sacramento-San Joaquin Delta. In Nonequilibrium Systems in Natural Waters. ACS, Washington, D.C. p. 131-180.

Fox, R.F. 1971. Entropy reduction in open systems. Journal Theoretical Biology, 31:43.

Hill, J., IV. 1973. Component description and analysis of environmental systems. Masters Thesis, Utah State University, Logan, Utah.

Karnopp, D., and R.C. Rosenberg. 1968. Analysis and simulation of multiport systems. M.I.T. Press, Cambridge, Mass.

Katchalsky, A., and P.F. Curran. 1965. Non-equilibrium thermodynamics in biophysics. Harvard University Press, Cambridge.

Martens, H.R., and O.R. Allen. 1969. Introduction to systems theory. C.E. Merrill Publishing Co., Columbus. 690 p.

Morowitz, H.J. 1968. Energy flow in biology. Academic Press, New York. 179 p.

Odum, E.P. 1971. Fundamentals of ecology. Saunders, Philadelphia. 574 p.

Ogata, K. 1967. State space analysis of control systems. Prentice-Hall, Englewood Cliffs. 596 p.

Patten, B.C. 1972. Personal communication. N.S.F. Institute, Utah State University, Logan, Utah.

Porcella, D.B., J.S. Kumagai, and E.J. Middlebrooks. 1970. Biological effects on sediment-water nutrient interchange. Proceedings ASCE, 96(SA4):911-926.

Porcella, D.B., K.L. Schmalz, and W.A. Luce. 1972. Sediment-water nutrient interchange in eutrophic lakes. Seminar on Eutrophication and Biostimulation, California Department of Water Resources, Sacramento, California. p. 83-109.

Porcella, D.B., V.D. Adams, P.A. Cowan, W. Holmes, S. Austrheim-Smith, J. Hill, IV, and E.J. Middlebrooks. 1973. Biological effects on interchange of metals and nutrients between sediments and water. OWRR Project Report. Utah Water Research Laboratory, Logan, Utah.

Prigogine, I. 1961. Introduction to thermodynamics of irreversible processes. Interscience Pub., New York.

Stephenson, R.E. 1971. Computer simulation for engineers. Harcourt Brace Jovanovich, Inc., New York. 504 p.

Wagman, D.D. et al. 1965. Selected values of chemical thermodynamic properties. Technical Note No. 270-1 N.B.S., U.S. Dept. of Commerce, Washington, D.C.

Walters, C.J. 1971. Systems ecology: The systems approach and mathematical models in ecology. In Fundamentals of Ecology, by E.P. Odum. Saunders, Philadelphia. 574 p.

Weast, R.C. (ed.). 1970. Handbook of chemistry and physics. The Chemical Rubber Co., Cleveland, Ohio.

PREDICTING THE EFFECTS OF NUTRIENT DIVERSION

ON LAKE RECOVERY

M. W. Lorenzen[1]

The role of nutrients in the eutrophication process has received widespread attention. A general conclusion has been that gross nutrient discharges should be halted. As has been pointed out by Fuhs (1972), regulatory agencies require predictions of the impact of present and future nutrient inputs on lake quality over a period of years. However, the effects of decreasing or eliminating nutrient inputs to a lake are not clear. The diversion of sewage from Lake Washington was dramatically successful in decreasing lake phosphorus and algal biomass, (Edmondson, 1972), yet diversion of 40 percent of the external phosphorus loading from Lake Sammamish has had no appreciable effect on either phosphorus levels or algal concentrations (Emery et al., 1973).

It is the purpose of this paper to summarize model derivations and show how different assumptions about nutrient exchange processes can result in varying predictions concerning lake recovery. Three different models will be considered. Model I is essentially that of Vollenweider (1969), in which the net loss of nutrient to the sediment on an annual basis is considered proportional to the average concentration of that nutrient in the water. Similar formulations have been used by Emery et al. (1973) and Megard (1971). Model II considers both loss to the sediments and return from the sediments as independent processes and assumes that the nutrient concentration in the sediment does not change appreciably over time. Model III considers coupled equations describing transfer into and out of the sediments.

All three models are idealized and simplified representations of complex processes. However, Models II and III not only represent improvement over existing models but also point out the important variables, identifying areas where future research should be concentrated.

---

[1]M. W. Lorenzen is with Battelle, Pacific Northwest Laboratories, Richland, Washington.

Model Derivations

Model I

Assuming that long term effects can be approximated by average annual values in a well-mixed lake, a mass balance can be written on the nutrient of interest. A diagrammatic representation of this case is shown below.

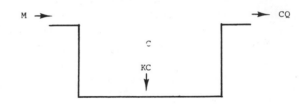

The rate of change in concentration with time is given by

$$\frac{dC}{dt} = \frac{M}{V} - \frac{CQ}{V} - \frac{KCA}{V} \qquad \ldots \ldots \ldots (1)$$

in which

M	=	mass flow in from all sources, g/yr
C	=	average annual nutrient concentration, g/m^3
K	=	net specific rate of loss to sediments, m/yr
Q	=	average annual outflow, m^3/yr
V	=	lake volume, m^3
A	=	surface area, m^2
C_o	=	nutrient concentration at time zero, g/m^3
t	=	time, yrs

Integration of Equation 1 from C_o at time zero to C at time t yields

$$C = \frac{M}{Q + KA} \left(1 - e^{-\frac{Q + KA}{V}t} \right) + C_o e^{-\frac{Q + KA}{V}t} \qquad . \ (2)$$

Equation 2 describes the concentration of nutrient in the water as a function of time as a result of changing the nutrient input rate, M. Evaluation of Equation 2 at infinite time yields an equilibrium concentration of

$$C_{eq} = \frac{M}{Q + KA} \quad \ldots \ldots \ldots \ldots \quad (3)$$

The time required to change from one equilibrium concentration C_o to within 10 percent of a new equilibrium value C_{eq} as a result of decreasing the nutrient input rate M is given by

$$t_{10} = \frac{V}{Q + KA} \left(2.3 + \ln \left[\frac{Q + KA}{M} C_o - 1 \right] \right) \quad . \quad (4)$$

A major deficiency of Model I is that the apparent rate constant K is determined from the *net* loss of nutrient to the sediments. The value of K is thus the difference between the rate into and the rate out of the sediments divided by the nutrient concentration in the water. Model II will show the different results obtained by treating the two processes independently.

Model II

A diagrammatic representation of a simplified lake in which nutrient fluxes into and out of the sediments are considered independently is shown below.

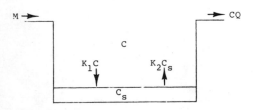

For cases in which the nutrient concentration in the sediments does not change significantly over time, a mass balance yields

$$\frac{dC}{dt} = \frac{M}{V} + \frac{K_2 C_s A}{V} - \frac{K_1 C A}{V} - \frac{CQ}{V} \quad \ldots \ldots \quad (5)$$

in which

M, C, Q, V, A, and C_o are as previously defined, and
K_1 = specific rate of nutrient transfer to sediment, m/yr
K_2 = specific rate of nutrient transfer from sediment, m/yr
C_s = nutrient concentration in sediment, g/m^3

The solution to Equation 5 is

$$C = \frac{M + K_2 C_s A}{Q + K_1 A} \left(1 - e^{-\frac{Q + K_1 A}{V} t} \right) + C_o e^{-\frac{Q + K_1 A}{V} t} . \quad (6)$$

In this case the specific rates K_1 and K_2 must be determined independently. The relationship between K of Model I and K_1 and K_2 is given by

$$K = \frac{K_1 C - K_2 C_s}{C} = K_1 - \frac{K_2 C_s}{C} \quad \ldots \ldots \quad (7)$$

The value of K is relatively easy to estimate for various lakes. It is possible that estimates of K_1 and K_2 could be obtained by plotting K versus C_s/C for various lakes or for the same lake at different points in time.

Evaluation of Equation 6 at infinite time yields the equilibrium concentration according to Model II. The concentration is given by

$$C_{eq} = \frac{M + K_2 C_s A}{Q + K_1 A} \quad \ldots \ldots \ldots \ldots \ldots \quad (8)$$

For Model II, the time required to change from one equilibrium concentration C_o to within 10 percent of a new value as a result of decreasing the nutrient input rate M is given by

$$t_{10} = \frac{V}{Q + K_1 A} \left(2.3 + \ln \left[\frac{C_o (Q + K_1 A)}{M + K_2 C_s A} - 1 \right] \right) \quad (9)$$

As will be shown later, the time required to reach a new equilibrium is relatively short for both Models I and II. However, the value of the new equilibrium concentration predicted by Models I and II can be quite different, depending on the relative values of M and $K_2 C_s A$.

Model III

To take into account the possibility that nutrients may be depleted from the sediments when nutrient input to the water is decreased, the value of C_s must be considered variable.

The diagrammatic representation of Model III is identical to that of Model II except that the coupled set of differential equations describing the nutrient concentration in the water and the sediments must be solved.

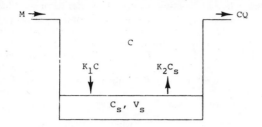

$$\frac{dC}{dt} = \frac{M}{V} + \frac{K_2 C_s A}{V} - \frac{K_1 CA}{V} - \frac{CQ}{V} \quad . \quad . \quad . \quad .(10)$$

$$\frac{dC_s}{dt} = \frac{K_1 CA}{V_s} - \frac{K_2 C_s A}{V_s} \quad . \quad . \quad . \quad . \quad . \quad . \quad .(11)$$

in which M, C, Q, V, A, C_o, K_1, K_2, and C_s are as previously defined and V_s = volume of sediment. The solution to the coupled set of Equations 10 and 11 is somewhat more complicated and the procedure will only be outlined here.

Let

$$X_1 = M/V \qquad g/m^3/yr$$

$$X_2 = -\frac{K_1 A}{V} - \frac{Q}{V} \qquad m/yr$$

$$X_3 = \frac{K_2 A}{V} \qquad m/yr$$

$$X_4 = \frac{K_1 A}{V_s} \qquad m/yr$$

$$X_5 = -\frac{K_2 A}{V_s} \qquad m/yr$$

then

$$\frac{dC}{dt} = X_1 + X_2 C + X_3 C_s \quad . \quad . \quad . \quad . \quad . \quad . \quad (10a)$$

$$\frac{dC_s}{dt} = X_4 C + X_5 C_s \quad . \quad . \quad . \quad . \quad . \quad . \quad (11a)$$

In matrix notation, Equations 10a and 11a can be written

$$[\overset{o}{C}] = [A] \, [C] + [D] \quad . \quad . \quad . \quad . \quad . \quad . \quad .(12)$$

in which

$$[\overset{o}{C}] = \begin{bmatrix} \dfrac{dC}{dt} \\[2mm] \dfrac{dC_s}{dt} \end{bmatrix}$$

$$[A] = \begin{bmatrix} X_2 & X_3 \\ X_4 & X_5 \end{bmatrix}$$

$$[C] = \begin{bmatrix} C \\ C_s \end{bmatrix}$$

and

$$[D] = \begin{bmatrix} X_1 \\ O \end{bmatrix}$$

The solution to Equation 12 is given by

$$[C] = e^{[A]t}[C_o] + \left(e^{[A]t} - [I] \right) [A]^{-1} [D] \quad .(13)$$

in which

$$[C_o] = \begin{bmatrix} C \\ C_s \end{bmatrix} \text{ at time zero } = \begin{bmatrix} C_o \\ C_{s,o} \end{bmatrix}$$

The concentration of nutrient in the water as a function of time is thus given by

$$C = \left[\frac{\alpha + X_5}{\alpha - \beta} e^{-\alpha t} + \frac{\beta + X_5}{\beta - \alpha} e^{-\beta t} \right] C_o$$

$$+ \left[\frac{X_3}{\alpha - \beta} e^{-\alpha t} + \frac{X_3}{\beta - \alpha} e^{-\beta t} \right] C_{s,o}$$

$$+ \frac{X_1}{X_2 X_5 - X_3 X_4} \left[\left(\frac{\alpha + X_5}{\alpha + \beta} e^{-\alpha t} + \frac{\beta + X_5}{\beta - \alpha} e^{-\beta t} - 1 \right) X_5 \right.$$

$$\left. - \left(\frac{X_3 X_4}{\alpha - \beta} e^{-\alpha t} + \frac{X_3 X_4}{\beta - \alpha} e^{-\beta t} \right) \right] \quad . \quad . \quad . \quad .(14)$$

in which X_1, X_2, X_3, X_4, and X_5 are as previously defined and α and β are given by the solution to Equations 15 and 16.

$$\alpha + \beta = -(X_2 + X_5) \quad . \quad . \quad . \quad . \quad . \quad . \quad .(15)$$

$$\alpha\beta = (X_2 X_5 - X_3 X_4) \quad . \quad . \quad . \quad . \quad . \quad .(16)$$

207

The concentration of nutrient in the sediment is given by a similar evaluation of Equation 13.

Evaluation of Equation 14 at infinite time yields an equilibrium concentration of M/Q. This is the expected result as the nutrient concentration in the sediment would reach a level at which net exchange would be zero and the concentration of nutrient in the lake would be equal to the incoming concentration.

Comparison of Models

Because the data available for comparing these models are very limited, hypothetical conditions have been used for example calculations. Table 1 summarizes the various parameters for four such lakes.

These values were used to calculate nutrient concentrations as a function of time for each lake, using each model. Plots of the results are shown on Figure 1. These figures reveal that the different models can predict very different results.

Discussion

These calculations emphasize that predicted results can vary substantially, depending on the model employed. The differing predictions occur even though the lake's characteristics have been precisely (although somewhat arbitrarily) defined.

Although all of the models are idealizations of reality, Model III is more complete than the others and

warrants further investigation. The writer has not yet had the opportunity to completely evaluate field data from lakes where nutrient diversion has been effected. However, preliminary calculations show that if a term $(-K_3C/V_s)$ is added to Equation 11 to account for some permanent nutrient loss to the sediment, the model can be calibrated to fit both Lake Washington and Lake Sammamish.

An additional factor which is seldom considered in lake recovery calculations is related to pre-diversion conditions. Any model should first be checked to insure that it accurately predicts pre-diversion lake conditions before the model is used to evaluate the effects of nutrient diversion. Figure 2 illustrates the importance of pre-diversion modeling.

This figure was constructed by using Model III to calculate the nutrient concentration for two different loading rates in a lake with the physical characteristics of lake 1 (Table 1), except that initial nutrient conditions were zero. The two curves plotted describe the time course of nutrient increase in a pristine lake if it were loaded at two different rates. If the lake were initially loaded at M = 100 x 10^6 g/yr and then the loading rate was reduced to 50 x 10^6 g/yr the results would depend on what point in time diversion took place. Case A illustrates that it is possible to reduce nutrient loading by a factor of two and still show an increase in average nutrient concentration in the lake. Case B shows a reduction in average nutrient concentration as a result of a reduction in loading rate. The dashed lines in the figure indicate that concentration as a function of time using Model III with different initial nutrient concentrations. These seemingly disparate results are due to initial conditions which are not at steady state with loading rates.

Table 1. Summary of parameters used in model calculations.

| | Lake | | | |
	1	2	3	4
M pre. div., g/yr	100x10^6	20x10^6	12x10^6	250x10^6
M post. div., g/yr	50x10^6	10x10^6	4x10^6	75x10^6
τ, yrs	3.5	3.5	3.5	3.2
Q, m^3/yr	857x10^6	86x10^6	86x10^6	900x10^6
V, m^3	3000x10^6	300x10^6	300x10^6	2884x10^6
A, m^2	100x10^6	20x10^6	20x10^6	87.6x10^6
depth, m	30	15	15	30
K, m/yr	8.4	6.0	6.0	26
K_1, m/yr	40.0	40.0	40.0	40
K_2, m/yr	0.0085	0.0085	0.0085	0.0085
C_s', g/m^3	200	100	100	100
V_s, m^3	10.0x10^6	2.0x10^6	2x10^6	8.7x10^6

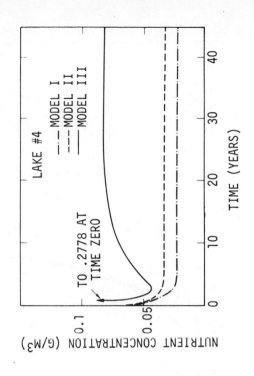

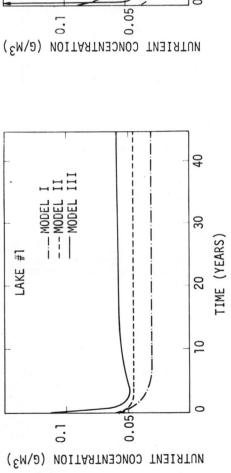

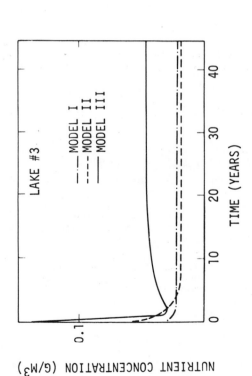

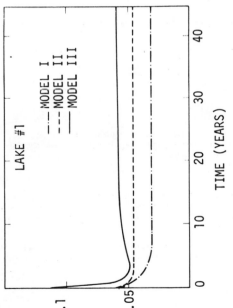

Figure 1. Calculated nutrient concentration for conditions summarized in Table 1.

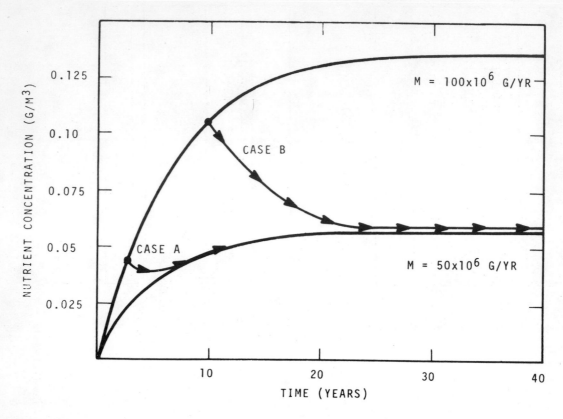

Figure 2. Nutrient concentration as a function of time calculated with Model III for a lake with the characteristics of lake 1 and no initial nutrients.

References

Edmondson, W. T. 1972. Nutrients and phytoplankton in Lake Washington. *In* Nutrients and Eutrophication. G. E. Likens (ed.), American Society of Limnology and Oceanography, Lawrence, Kansas, pp. 172-193.

Emery, R. M., C. E. Moon, and E. B. Welch. 1973. Delayed response of a mesotrophic lake after nutrient diversion. Journal Water Pollution Control Federation 45:913-925, May.

Fuhs, G. W., S. E. Demmerle, E. Caneli, and M. Chen. 1972. Characterization of phosphorus limited plankton algae. *In* Nutrients and Eutrophication. G. E. Likens (ed.), American Society of Limnology and Oceanography, Lawrence, Kansas, pp. 113-133.

Megard, R. O. 1971. Eutrophication and the phosphorus balance of lakes. Presented at the 1971 Winter Meeting of the American Society of Agricultural Engineering, Chicago, Illinois, December.

Vollenweider, R. A. 1969. Possibilities and limits of elementary models concerning the budget of substances in lakes. Arch. Hydrolobiology, 66:1-36, April.

DISCUSSION

G. D. Cooke[1]

The eutrophication process has been discussed here from two levels of biological organization, the cell and the ecosystem. These points of view of course reflect the various interests and backgrounds of the conferees. Nevertheless, I believe that to model a process requires a definition of that process and, at least for ecology, some agreement on the actual level of biological organization involved in the process.

Eutrophication, I believe, is a process in which nutrient cycles of watershed ecosystems become opened through agriculture and urbanization so that conserved nutrients, as well as those imported by people (or in some cases, by industrial processes), are added to aquatic communities on the watershed at rates sufficient to significantly increase, on a long-term basis, their availability to autotrophs. The *process* is therefore one which takes place at the ecosystem level of organization wherein terrestrial events exert a dominating influence on the metabolism of the aquatic communities of the watershed.

Justifiably aquatic ecologists are under some public pressure to not only understand this process and its causes but also to give some meaningful estimates of methods and costs of restoration and the length of time for recovery. The boater, swimmer, fisherman, and home-owner accurately see the process and response to be not only sewage import and nuisance blooms of algae, but also, and perhaps of equal importance, the results of import and the acceleration of certain interactions between trophic levels and lake zones which together result in altered fishing, vast beds of macrophytes, "mucky" sediments, bacterial blooms, poor boating, smell, loss of property values, etc.

I believe that models that deal with such processes as uptake kinetics will be of some considerable significance to our fundamental understanding of aspects of the process, and particularly the response of species populations to certain changes in their environments. It seems however, that there is reason to keep clear that the eutrophication process is first an ecosystem level one and that any hope of achieving restoration and effective public understanding of the consequences of driving watersheds towards developmental stages will ultimately come mainly through holistic models which emphasize this level of organization, the change in material cycling during eutrophication, and the response and interactions of trophic levels and lake zones.

Perhaps the restoration process can be described as the reverse of the eutrophication process, and some guidelines for long-term watershed planning and management can be developed from this point of view.

[1] G. D. Cooke is with the Department of Biological Science, Center for Urban Regionalism, Kent State University, Kent, Ohio.

DISCUSSION

A Note on Fundamental Modeling Principles

and Algal Limiting Nutrients

F. H. Verhoff[1]

There are only three fundamental concepts employed in the construction of quantitative models of nutrients cycling to aquatic environments as well as used in the qualitative understanding of what occurs therein. The first principle and probably the most rudimentary is the conservation of mass; the Streeter-Phelps equation is the conservation of oxygen mass. Secondly, the rate at which certain processes occur must be quantized. In other words, the kinetics of oxygen transfer from the atmosphere to the water body, a mass transfer operation, and the rate of oxygen utilization by the bacteria, a chemical transformation, must be known for the formulation of the Streeter-Phelps equation. Whenever the conservation of more than one chemical species is considered in a model, then the stoichiometric relationship between these two entities during chemical transformations are specified. For example, if along with the conservation of oxygen, the conservation of carbon dioxide was modeled, then the ratio of oxygen consumed to carbon dioxide generated by the bacteria must be specified. Similar stoichiometric relationships are necessary to understand the ratios of various nutrients consumed during algal growth. Thus the three principles used in the comprehension of nutrient cycling phenomena of aquatic environments are conservation of mass, kinetics of interactions, and stoichiometry of chemical transformations.

These principles are used in two different approaches to specify what the "limiting" algal nutrient is in a given lake; under certain conditions these two approaches do not identify the same nutrient as "limiting." According to one conceptual approach, there can be only one "limiting" nutrient whereas the other method of reasoning states that more than one nutrient can be "limiting" simultaneously. Other conceptual difficulties and misunderstandings have resulted from the simultaneous use of two methods for determining the same quantity, the "Limiting Nutrient."

The stoichiometric limiting nutrient uses the principles of conservation of mass and the stoichiometric composition of the algae to determine which nutrient will ultimately limit the algal biomass if that one nutrient is totally depleted. In order to calculate the stoichiometric limiting nutrient, one divides the concentrations of each available nutrient in the water by the stoichiometric requirement of algal for that nutrient, and states that the nutrient generating the lowest ratio is the "limiting nutrient." The nutrients usually associated with stoichiometric limitations are carbon, nitrogen, and phosphorus. Using this process, there can be only one limiting nutrient at any given time.

In contrast to the stoichiometric limiting nutrient, there also exists the kinetic limiting nutrient or nutrients. At any given instant in time, a nutrient is considered kinetically limiting for the growth of algae, if an increase in the water concentration of this nutrient results in a higher growth rate for the algae. In many cases, more than one nutrient satisfies this criteria, and thus, often more than one nutrient is kinetically limiting. Many substances can be kinetically limiting including the macronutrients, the micronutrients, and the cations Na and K. Frequently, it is nearly impossible to separate the kinetic importance of various nutrients, e.g., if NaH_2PO_4 is added to an algal culture, either the sodium or the phosphate could simulate growth.

In summary, caution should be exercised in the specification of the "limiting nutrient," i.e., it should be clearly stated whether the limiting nutrient is stoichiometric or kinetic. From this discussion, it might be concluded that for a more precise determination of the limiting nutrient or nutrients all three fundamental principles should be considered.

---

[1] F. H. Verhoff is with the Department of Chemical Engineering, University of Notre Dame, Notre Dame, Indiana.

LIST OF PARTICIPANTS

Name	Affiliation	Location
Dean Adams	Utah State University	Logan, Utah
David A. Bella	Oregon State University	Corvallis, Oregon
Victor J. Bierman, Jr.	University of Notre Dame	Notre Dame, Indiana
Jay A. Bloomfield	Rensselaer Freshwater R.P.I.	Troy, New York
Patrick L. Brezonik	University of Florida	Gainesville, Florida
Lawrence A. Burns	University of Florida	Gainesville, Florida
G. Dennis Cooke	Kent State University	Kent, Ohio
Peter Cowan	Utah State University	Logan, Utah
James W. Falco	U.S. Army Corps of Engineers	Vicksburg, Mississippi
G. Wolfgang Fuhs	N.Y. State Dept. of Health	Albany, New York
William J. Grenney	Utah State University	Logan, Utah
Fred Hagius	Utah State University	Logan, Utah
James Hill	University of Georgia	Athens, Georgia
Dale D. Huff	University of Wisconsin	Madison, Wisconsin
Eugene Israelsen	Utah State University	Logan, Utah
Walter R. Ivarson	University of Wisconsin	Madison, Wisconsin
Norbert A. Jaworski	E.P.A.	Corvallis, Oregon
Joseph F. Koonce	University of Wisconsin	Madison, Wisconsin
D. P. Larsen	E.P.A.	Corvallis, Oregon
Ray R. Lassiter	E.P.A.	Athens, Georgia
G. Fred Lee	Texas A&M	College Station, Texas
Marc W. Lorenzen	Battelle - Northwest	Richland, Washington
Thomas E. Maloney	Pacific N.W. Env. Research Lab	Corvallis, Oregon
Kenneth W. Malueg	Pacific N.W. Env. Research Lab	Corvallis, Oregon
P. H. McGauhey	University of Calif.	El Cerrito, Calif.
E. Joe Middlebrooks	Utah State University	Logan, Utah
G. Wayne Minshall	Idaho State University	Pocatello, Idaho
Richard A. Parker	Washington State University	Pullman, Washington
Bernard C. Patten	University of Georgia	Athens, Georgia
Spencer A. Peterson	Pacific N.W. Env. Research Lab	Corvallis, Oregon
Donald B. Porcella	Utah State University	Logan, Utah
Bill Richardson	E.P.A.	Grosseile, Michigan
Walter M. Sanders III	E.P.A.	Athens, Georgia

Name	Affiliation	Location
Frieda B. Taub	University of Washington	Seattle, Washington
Bruce A. Tichenor	E.P.A.	Corvallis, Oregon
Paul D. Uttormark	University of Wisconsin	Madison, Wisconsin
Frank Verhoff	University of Notre Dame	Notre Dame, Indiana
Gene Welch	University of Washington	Seattle, Washington
J. David Yount	E.P.A.	Washington, D.C.

INDEXES

AUTHOR INDEX

A

Abrams, A., 107
Adams, M., 55
Adams, M.S., 148, 150
Adams, V.D., 203
Adinarayana, A., 108
Allen, O.R., 189, 191, 203
Allen, T.F.S., 50, 54, 148
Allison, D.E., 13
Andersen, J.C., 186
Anderson, G.C., 13
Andersson, G., 5, 13
Andrews, J.F., 138
Aoki, S., 108
Armstrong, D.E., 52, 54
Armstrong, F.A.J., 108, 186
Asplund, O.W., 186
Astor, P.H., 85, 86, 193, 203
Austrheim-Smith, S., 203
Azad, H.S., 90, 93, 97, 98, 106

B

Baarda, J.R., 91, 107
Bachmann, R.W., 82
Bacon, M.L., 164, 169
Ba Hli, F., 189, 203
Bannerman, R.T., 54
Barber, J., 98, 106
Barrett, J., 108
Bella, D.A., 107
Bertalanffy, L. von, 193, 194, 203
Bhargava, T.N., 57, 60, 72
Bierman, V.J., Jr., 89, 109
Bishop, A.B., 171, 186
Black, C.A., 72
Bloom, S.G., 111, 120
Bloomfield, J.A., 55, 139, 148, 150
Boormen, W., 13
Booth, R.S., 55, 150
Borchardt, J.S., 90, 93, 97, 98, 106
Boudart, M., 188, 189, 203
Bovard, J., 13
Brezenski, F.T., 74, 82
Brezonik, P., 17, 29, 31
Brown, R.L., 94, 106
Brunskill, G.J., 108, 186
Brylinski, M., 85, 86, 87
Bryson, R.A., 75, 82
Buckingham, R.A., 138
Buller, R., 64, 71
Burkhead, C.E., 114, 120
Burris, R.H., 108
Bush, R.M., 7, 13, 98, 106

C

Cale, W.G., 85, 87
Callen, H.B., 188, 189, 203
Canelli, E., 31, 107, 210
Caperon, J., 90, 102, 107, 134, 138
Carberry, J.B., 91, 107, 109
Carpenter, E.J., 102, 107
Castenholz, R.W., 97, 107
Chapman, R.N., 111, 120
Chamberlain, W., 108
Chen, C.W., 89, 107, 111, 120, 134, 138
Chen, M., 107, 210
Chuck, F.J., 169
Churchill, M.A., 132, 138
Clarkson, S.P., 87
Clesceri, L.S., 55
Clesceri, N.L., 150
Close, C., 189, 203
Coatsworth, J.L., 19, 30
Colon, E.M., 150
Colon, H.M., 55
Commoner, B., 187, 203
Confer, J.L., 70, 72
Cooke, G.D., 57, 58, 72, 211
Cooney, C.L., 120
Cordeiro, C.F., 107, 109, 111, 120
Corwin, N., 74, 82
Cowan, P.A., 203
Crawford, A.B., 186
Crawford, N.H., 36, 54
Cronberg, G., 13
Cullen, R.S., 36, 54
Curl, H. C., Jr., 107
Curran, P.F., 188, 203

D

Davis, R.A., Jr., 146, 148
DeGroot, S.R., 188, 189, 203
Demmerle, S.D., 31, 107
Demmerle, S.F., 210
DePinto, J.V., 107
Derse, P.H., 108
DeRusso, R.M., 189, 192, 193, 203
Desoer, C.A., 83, 87, 189, 203
Dettmann, E.H., 33, 52, 54, 146, 148, 150
DiCesare, F., 148
Dingman, S.L., 68, 72
DiToro, D.M., 19, 30, 89, 107, 109, 111, 120, 121, 130, 138, 203
Dodson, A.N., 121, 130
Dugan, G.L., 186
Dugdale, R.C., 19, 30, 121, 130
Dutton, J.A., 75, 82

SUBJECT INDEX

225